SUBMARINER

ROLEX

SPECIAL EDITION WRISTWATCHES

SCHIFFER PUBLISHING
4880 Lower Valley Road • Atglen, PA 19310

ROLEX

Rolex Special Edition Wristwatches

Editorial

Hans Wilsdorf,
father of the modern wristwatch

The presentation of the new Submariner was not the only reason for issuing this special edition, but it certainly was an important added factor. With this diving watch, Rolex has again come one step closer to the ideal of perfection. With a diameter of 41 millimeters (mm) and an in-house watch movement of the newest generation, equipped with a scratch-resistant ceramic bezel and an adjustable link bracelet that is as robust as it is comfortable, this modern sports watch really leaves nothing to be desired. And so, on the first day of sales, long lines formed in front of the concessionaires, because every watch enthusiast wanted to try his or her luck to get one of the first specimens—or at least to make sure of the option to get one of the next series. At the same time, the untrained eye can barely detect the changes in comparison to the previous model—itself by no means any kind of slow seller.

Hans Wilsdorf recognized early on what watch enthusiasts really want. It is not every year that a flood of new models, which had already been touted in the spring, are already old hat when they finally come on the market at Christmas simply because of the mere prospect of the upcoming innovations. No overloaded dials, where you can hardly read everything that nobody needs to know about all the time anyway. No delicate mechanisms to control the hands that playfully dance back and forth, and which are put off their stride permanently by an unintentional bump against the edge of a table, which in turn results in expensive repairs.

First and foremost, as the entrepreneur from Kulmbach, in Upper Franconia, might have thought, a wristwatch must be able to display the time, and in fact do so at any point in time, under the most-adverse circumstances, and with absolute precision. First, he made his watchcases waterproof, because moisture and dust could damage the movement. He screwed down the crown, like a small bell with a seal, into a tube on the watchcase. To ensure that it would not be necessary to loosen this hermetic screw connection every day, he equipped his watch movements with a self-winding mechanism with an oscillating rotor. With the creation of the Oyster Perpetual, this meant that the modern wristwatch was, to a large extent, fully developed.

That was in 1931.

Wilsdorf left his successors a foundation that tied the capital up within the company, and they thanked him by making wise decisions to verticalize the production facilities and by taking carefully considered steps to consolidate the model range. And so, for more than sixty years Rolex has been positioned to be crisis-proof with its small collection of three-hand watches—some sports watches, some elegant—an iconic chronograph and a handful of interesting complications.

A benchmark for the Swiss watch industry and at the same time in a class of its own.

Peter Braun

ROLEX
OYSTER PERPETUAL
SUPERLATIVE CHRONOMETER
OFFICIALLY CERTIFIED
COSMOGRAPH

Table of Contents

I.
The History of the Rolex Brand

At the Sign of the Crown

The French structural sociologist Claude Levi-Strauss is the author of the sentence "Myths are hollow and round." Which is to say: If you break the outer shell, you end up holding nothing in your hand. Perhaps even more apt is the image of the onion, with nothing left over after the last layer is peeled away. Myths about companies and their products are also layers and shells from history, and history in turn emerges from what is consciously staged, but also from what happens by chance, and gains its significance only in retrospect. The mythos of a brand and its products unfolds itself only from this accumulation, meaningful networking, and the knowledge of the distinction of the events. Especially in the age of increasingly extensive industrial production, it is the brand that allows the consumer to orient oneself and to differentiate.

Hans Wilsdorf,
founder of the Rolex watch brand,
at a young age

A NAME BECOMES A CONCEPT

There is no doubt that, among all the iconic brand names that were lined up to make their triumphal advance at the beginning of the twentieth century, at the top is Rolex, an artificial name registered in 1908. A name invented by the German watch manufacturer Hans Wilsdorf (1881–1960), who came from the German city of Kulmbach, which is said to be derived from the French, *horlogerie exquise*. However, this has not been proven.

The Rolex company founder set himself the goal of presenting a precise wristwatch that would be able to compete with the pocket watches that were still dominant at the time. He was also the one who defined the fundamental values of the brand that still define the essence of the brand today: the highest quality of craftsmanship in manufacturing and workmanship, the highest possible level of precision possible, plus being robust and suitable for everyday use—right through to models for a range of purposes that perform under extreme conditions.

For the Rolex mythos, Wilsdorf plays a role similar to that which Enzo Ferrari did for the automobile brand of the same name: one of them wanted to present the best racing car to the world; the other the most precise and most durable wristwatch.

Wilsdorf was more than a watch technician and watchmaker; he was a gifted salesman, while also being a marketing expert—a craft that at that time was still carrying on its trade under the label of "propaganda." He rapidly realized how to use people as "testimonials" to convey the product's message. With tremendous willpower and determination, provided with a sure instinct, he put his ideas and ideals into effect. At the same time, his biography began with a severe stroke of fate: At age twelve he lost both parents in quick succession: first his mother, then his father. His relatives then took care of him and his two siblings; they sold the father's business and invested the proceeds profitably.

After his time at boarding school in Coburg, Germany, as well as passing his Abitur—the German university-qualifying examinations—he completed a commercial apprenticeship with a man in Bayreuth who traded in artificial pearls all over the world. Commercial work suited him; at the same time, he was very interested in watch technology and foreign languages. So, when he was just barely twenty years old, he went to La Chaux-de-Fonds in Switzerland to work for a large watch export company. For eighty Swiss francs a month, he handled the English correspondence, did office work, and every day wound the pocket watches in which the company traded in.

The precision of keeping time rapidly became his obsession. Using part of his inheritance from his father, he bought three gold pocket watches and had their accuracy recorded at an observatory in rate certificates. Afterward he sold the watches for a profit.

Before moving to London—which was then the center of the industrial world—in 1903, Wilsdorf completed his one-year military service in the Imperial Army of the German Reich. In 1905, together with the much-older Alfred James Davis, he founded his own watch-trading company in London under the name "Wilsdorf & Davis." He had to borrow some of his capital from his brother and sister, because his 30,000 gold marks from his inheritance from his father had been stolen while on his passage to England.

If it had been the pocket watch that awakened his interest and enthusiasm for the world of timekeeping, now above all it was the wristwatch—the novelty in this manufacturing sector—and its rapidly developing market.

At the Aegler clockwork factory in Biel, Switzerland, he bought so many high-quality small-lever movements that the sum that had to be paid came to five times the amount of the company's capital. But his plan worked, and success proved him to be right.

By 1907, the flourishing company opened a branch in La Chaux-de-Fonds, Switzerland. By 1908, Wilsdorf & Davis was one of the largest companies in the European watch trade and had two hundred models in its range. The watches were sold anonymously, or under the logo of the respective dealer. Only the watchcases were stamped with "W/D," for Wilsdorf & Davis.

This displeased the man in charge, as did the fact that at that time, wristwatches were still almost exclusively reserved for ladies' wear and were considered to be "unmanly."

Extreme woman athlete Mercedes Gleitze swam across the English Channel wearing a Rolex, which she probably did not wear on her arm but, rather, around her neck.

Oyster Viceroy (1940) in a two-tone watchcase, Prince Brancard "Observatory Quality" chronometer (1936), and an Oyster "Channel Swimmer" Watch (1935)

Daily Mail

The Marvellous
ROLEX
WRIST WATCH
The World's Best by Every Test

MAKE IT A ROLEXMAS

ROLEX OYSTER

FREE!

THE ROLEX WATCH CO. LTD. GENEVA SWITZERLAND

With an automatic self-winding mechanism and a watertight screw-down crown, the "Oyster" was perfect from 1931 onward. Shown in the picture above, an Oyster Perpetual Chronometer dating from 1939.

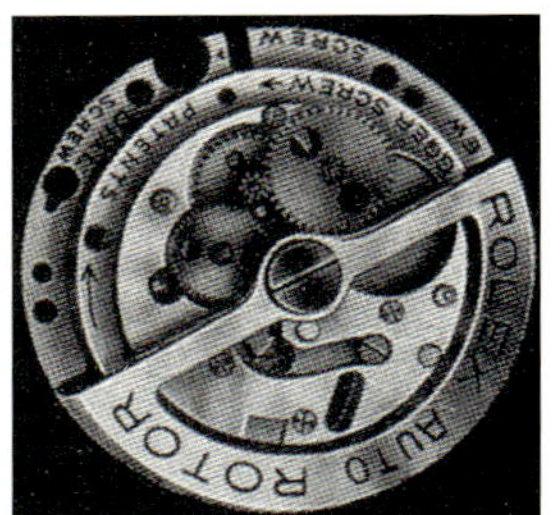

AGAINST ALL ODDS

As a first step, he came up with a product name. As he wrote on the subject in his working handbook: "It was so short and at the same time it was so compelling that there was enough space on the dial for the name of the English dealer next to it. But what was particularly valuable: Rolex has a good sound, is easy to remember, and is also pronounced the same way in all the European languages."

It would take twenty years for the new name to become established. Initially, Wilsdorf resorted to a trick: in the boxes, which contained six pieces, only two watches at a time were labeled "Rolex," and then later it was three or four, and thus the name also caught on in the dealers' show windows.

Of course, the young company had to provide a proof of quality for these small watch movements. Would it be possible for the delicate ladies' watches to stand up to the respectable chronometers in the men's vest pockets in terms of precision? They could!

As early as 1910, Wilsdorf had obtained a first-class rate certificate in Biel for a wristwatch movement with a diameter of 24.81 mm. In 1914, Rolex achieved the feat at the Kew Observatory in England: it was there that the first wristwatch was honored with a class A rate certificate because it had achieved the same timekeeping performance as a marine chronometer. All of a sudden Wilsdorf belonged to the circle of the most-renowned watchmakers in England—and achieved that with a wristwatch!

From the start, the watch movements were supplied by the Aegler company in Biel, behind which stood Jan Aegler and his company, founded in 1878. Since 1881, this company has been based in Rebberg, a district of Biel city, and had been exporting ladies' wristwatches all over the world since 1900—with the exception of England (starting in 1913), so as not to compete with its important business partner Rolex. In 1914, the company was converted into a stock corporation under the name "Aegler SA, Rolex Watch Company." It employed two hundred people and was the exclusive supplier for the "Wilsdorf & Davis Rolex Watch Company." Aegler remained independent as a company—a situation that would change only in 2004, when Rolex president Patrick Heiniger bought the company for the proud sum of 2.5 billion and then incorporated it into Rolex SA.

WAR AS THE FATHER OF ALL THINGS

"Bellum omnium rerum pater est," so runs the Latin phrase, and in relation to the wristwatch this may indeed be true. In the course of the many English colonial wars and not least the First World War, the watch moved to the wrist—to the place where it could be read quickly and easily.

This development brought Rolex the business success it desired, even as the end came for the Wilsdorf & Davis Rolex Watch Company in 1919, when English import duties rose to 33.3 percent. Export activities were transferred to the Biel office, and Wilsdorf himself moved to Geneva with his wife. In 1920, the company was renamed "Montres Rolex SA" after Wilsdorf severed ties from his not-well-loved partner. The movements continued to be manufactured in Biel, while the cases were made and the watches were assembled in Geneva. In 1925, the brand name was enhanced by a brand symbol in the form of a five-point crown, which has adorned all Rolex watches from 1939 to the present day.

After Wilsdorf's death in 1963, André J. Heiniger took over the presidency of Montres Rolex SA and the Hans Wilsdorf Foundation. The picture at left shows a Rolex Oyster Datejust Chronometer from 1952.

The history of the company is marked by the tenacious struggle to perfect the wristwatch, something that has to be constantly reconfirmed by official test certificates. The observatories of Kew, Geneva, and Besançon became the important places to go, and in essence Rolex has not abandoned this policy of making an independently tested watch up to the present. Rolex still remains the watch manufacturer with the most certified chronometers in the world. As early as 1968, Rolex had reached the million mark for its chronometers, and to date there may well be more than thirty million certified Rolex watches altogether. And almost a million chronometers are added every year.

HERMETICALLY SEALED

Water—that is, moisture—has been a natural enemy of the mechanical watch movement from the start. Basically, there never had been a water-resistant pocket watch, and it would take until the 1920s before a case design appeared on the market that made it possible to hermetically protect the movement from external influences. Rolex was an important pioneer in this development. In addition to mechanical robustness and achieving high-level chronometric performance, making a water-resistant watch was Wilsdorf's declared aim. This was made reality by means of the carefully sealed case parts that are screwed down against each other, a crown with screw thread and seal, and a positive-fitting crystal. A name was quickly found for it: the "Oyster"—the oyster as a symbol of the hermetic seal.

The patent for the first "Oyster" was granted in 1926, making it possible to present this innovation to the astonished public. When the young English stenographer and typist Mercedes Gleitze swam across the English Channel in fifteen and a half hours on October 27, 1927, Wilsdorf became aware of the young extreme athlete. When she took on a second crossing at the end of October, she was wearing a Rolex Oyster. Her achievement made it clear to the entire world that the breakthrough of a water-resistant wristwatch had been made. Although Mercedes had to give up after about ten hours, when she was 11 kilometers from the coast, because the cold was just too great, it was still a success that a wristwatch had been able to withstand the elements for so long. This triumph was announced in a full-page advertisement on the front page of the *British Daily Mail* on November 24, 1927, which cost Wilsdorf 40,000 Swiss francs. Suddenly everyone was talking about the Rolex, "the wonder watch that defies the elements." For another advertising ploy, Rolex used small aquariums that the concessionaires used to display the watches—submerged and surrounded by swimming goldfish—to astonished window shoppers.

However, it quickly became apparent that the screw-down crown was the weak point of the design, because it had to be opened and closed every day for winding the watch; this meant that it was subject to considerable wear and tear. In 1931, Rolex developed an automatic winding mechanism with a pivoting oscillating weight or rotor. This was patented in 1933, and you can justifiably claim that this revolutionized the development of the wristwatch—although only when the patent protection had expired and other manufacturers adopted the system.

WOUND AUTOMATICALLY

The Oyster with reference number 1858 was the first Rolex with a rotor mechanism. After being worn for six hours, the advertisement promised, the movement would be fully wound up. However, it didn't make this promise without adding that the watch could also be provided with energy by winding the crown manually as before.

In 1945, series production of the date function started, which would abruptly change at midnight. This was the birth hour of the "Datejust." This same year, the 50,000th chronometer was certified in Biel.

Rolex started out on its way into the twenty-first century with innovative technology, a conservative model policy, and restructured manufacturing facilities.

Patrick Heiniger took over the chairmanship of Montres Rolex SA in 1992 and set the course for the complete verticalization of production.

The chronology of the company's achievements can still be found today on every Rolex watch dial. "Superlative Chronometer Officially Certified" recalls the strict test criteria, "Oyster" the water-resistant case design, "Perpetual" the automatic self-winding system developed by Rolex, and "Datejust" the date that springs forward exactly at midnight. The compressed history of a long technical development is documented in the smallest space possible.

Rolex also made watches that are luxurious and elegant and are not designed for sports wear: the "Prince," a rectangular watch with a manual-winding movement and a decentralized small seconds that came on the market at the end of 1928, aimed at the world of the rich and beautiful—"The Watch for Men of Distinction," according to contemporaneous advertising. Some four hundred watches of this type were ordered for the silver jubilee of King George V, of course also as certified chronometers.

The model was so successful that it remained in the range for forty years and recently has even experienced a revival—this, too, is a hallmark of this company, which certainly does not unsettle its customers by making hectic model changes. Meanwhile, it is the "Submariner" that now holds the internal brand record at Rolex; they have been making this watch for almost seventy years without even appearing to be the least bit antiquated. Imagine any car model being manufactured for so long: Mercedes-Benz would still have to be offering a modified form of the "gull wing" model from 1955. Actually, a delightful thought . . .

THE FOUNDATION MODEL

In 1944, Wilsdorf suffered a severe blow of fate: his wife, May Florence, died. Since the marriage had remained childless, he transferred his shares in Montres Rolex SA to the Hans Wilsdorf Foundation. This was a wise decision with regard to the company's continued existence, as has been proven by other companies that function as foundation models, such as the piston manufacturer Mahle, or the Bosch company, both based in Stuttgart, Germany. Considerable financial resources have flowed and continue to flow from all of these foundation model companies—as is also the case for Rolex—into scientific, charitable, or social projects all over the world.

The seventieth birthday of the well-loved man in charge was celebrated in 1951 with a four-day festival in Geneva—after all, it was also to commemorate his fiftieth year in the service of timekeeping. Moreover, this was also the anniversary year of twenty-five years of the Oyster watchcase and twenty years of the self-winding mechanism. Despite his advanced age, Wilsdorf continued to determine the fate of the company, even though he now had two other directors at his side. During the summer months he lived with his second wife on the south shore of Lake Geneva. Every morning, his chauffeur, Rüttimann, drove him to work in a Mercedes-Benz—a make of car to which he had been loyal since 1935.

When he died in 1960 at age seventy-nine, that was the end of an era. From 1963 on, his confidante André J. Heiniger, as president of Montres Rolex SA and the Hans-Wilsdorf Foundation, directed the company's fortunes until 1992, when his son Patrick took over the company leadership. Under his direction, Rolex began the systematic verticalization of all means of production and the integration of all its suppliers, which led to the expansion of watch movement production in Biel and the construction of a new manufacturing center in the Geneva suburb of Plan-les-Ouates.

EXPANSION AND INTERNATIONALIZATION

Until 1965, the staff were working at Montres Rolex SA in more than cramped conditions on three floors in the Rue du Marché, with some of the four hundred employees spread out among neighboring buildings. At that time there were around 250 people employed to make watch movements in Biel.

Ever since the Great Depression of the 1930s and the accompanying devaluation of the British pound had made the English market difficult terrain, Rolex has focused on internationalization. This was taken into account by its subsidiaries worldwide, which Hans Wilsdorf was still looking after personally in the 1950s.

Wilsdorf did not live to witness the company's move to its new building on Rue François-Dussaud in 1965, but he did witness the launch of the new Oyster models for professionals and specialists, which started in 1953 with the "Turn-O-Graph" and the Submariner diving watch. Hard-wearing sports utility watches for mountaineers, pilots, divers, oceanologists, racing drivers, scientists, and adventurers form the moral backbone of the brand to this day. On the other hand, these models cater to a rather idiosyncratic customer taste, including sports watches made of precious metals and set with diamonds.

Despite some models with a moon phase or full calendar, and even a split-seconds chronograph, Rolex remained true to its minimalist principle for more than a century. There are no tourbillons, no carillons, no perpetual calendars, and certainly no "Grande Complication" adorning the history of this watch brand. This has not harmed the brand's luster; quite the contrary.

With its few but well-thought-out models, Rolex has significantly shaped a century of wristwatch history.

Dr. Harry Niemann

Gian Riccardo Marini (*left*) followed the brief presidency of Bruno Meier (2008–2011) "ad interim," and in 2014 he handed over the reins to Jean-Frédéric Dufour (*right*).

II.
Robust, Waterproof, Self-Winding Automatic

The Long Way to an Elegant Sports Watch

Probably no watch design has shaped the image of the classic sports watch more than the Rolex Oyster. First and foremost comes the Submariner—but along with it the GMT-Master and the Explorer—which in their range of designs have shaped the new type of watch known as the "tool watch": high-quality watches for a wide range of requirement profiles that function reliably at enormous altitudes, at extremely low or absurdly high temperatures, at great water depths, or in very strong magnetic fields. Then there were the chronographs, which made it possible—in a world increasingly dominated by speed—to precisely record even the smallest unit of time.

Left: Rolex Oyster Perpetual "Bubble Back" with concealed lugs ("hooded lugs"). The large picture shows the "Milgauss," successfully relaunched in 2007 with magnetic-field resistance up to 80,000 A/m.

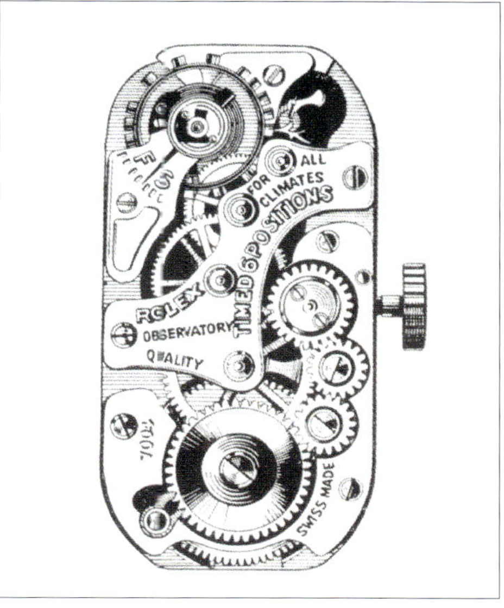

THE DEVELOPMENT OF THE SPORTS WATCH

The Rolex brand became popular at the end of the 1920s with its "Prince" model, an elegant rectangular watch with a two-part dial: the decentralized seconds hand featured a circular scale that was just as large as the time display with the hour and minute hands. From 1928 onward this watch was available in different case designs, including the "Railway" with a stepped case, or the "Brancard" with a tapered watchcase, and sometimes also in a range of combinations of precious metals. The dimensions of the eighteen-jewel shaped movement, caliber T.S. Reference 300, were 16.9 × 32.7 mm. It had a swan's-neck fine adjustment, was usually regulated at chronometer standard, and had a stable bearing bridge for the balance wheel. This distinguished it from the other "off the shelf" movements. Aegler, the watch movement maker, also supplied other manufacturers such as Gruen and Alpina—a circumstance that makes it easy for counterfeiters to produce new "Princes" today, and collectors should be correspondingly careful about any offers that are all too inexpensive. Apropos this: in 2005, Rolex presented a modern reinterpretation of this model, chronometer-certified with manual winding, but without market success.

While the early-form watches were equipped with manual-winding movements, as were the chronographs and the "pre-Oyster" models, the water-resistant "Oysters" were soon being made with the Perpetual caliber, which was manufactured exclusively for Rolex by Aegler in Biel; in 1933 a patent was granted for the caliber's automatic bidirectional rotor. These early self-winding models owe their posthumously

On the left, a Rolex Prince from 1935, with its characteristic dial design. *In the middle*, an early Prince manual-winding movement, regulated in six positions "for all climates." *On the right* is the new interpretation of the Prince from 2005, with a manual-winding movement; however, it wasn't a success on the market.

On his first ascent of Mt. Everest, Sir Edmund Hillary wore an Oyster like the one shown in the center picture. During the 1930s, Rolex developed and produced diving watches for the Italian luxury watch manufacturer Officine Panerai.

acquired nickname of "Bubble Back" to their bulbous case back, which created space for the rotor. The elegantly integrated lugs also set the new watches apart from other round models, which, by featuring soldered-on wire brackets to attach the watch strap, were still very reminiscent of converted pocket watches. A two-tone model in steel and gold was also offered for the first time when the "Hooded Bubble Back," with concealed lugs, came on the market in summer 1938. The typical Rolex metal bracelet, which forms an optical unit with the watch, also made its debut here. To this day it continues to have a lasting impact on the visual and tactile impression made by the sports watch line.

ROBUST AND ATHLETIC

As early as 1936, Rolex developed a water-resistant watch for the Florence-based company Panerai, designed for the Italian navy. The first prototypes of the "Radiomir" were even made at Rolex. Yet, the brand's great success story actually began in 1953, with the "Submariner" diving watch; this, however, was preceded by the "Turn-O-Graph," a "tool watch" with a rotating bezel, advertised by Rolex as a simple chronograph.

The Explorer was also introduced in 1953, followed in 1954 by the GMT-Master (with a second time zone and an additional twenty-four-hour hand), and by the "Milgauss," which had magnetic field resistance of up to 1,000 gauss. The "Day-Date," introduced in 1956, offers a display of the day of the week and the date; important information for any businessman. The simpler "Datejust," which has a date display that jumps forward exactly at midnight, was also available as a women's watch right from the start, and Rolex spared no effort to fulfill its ambition of having the

delicate movements also certified as chronometers—no easy undertaking!

In 1961, Rolex presented the Cosmograph Daytona; in 1971 the Sea-Dweller 2000 (water resistant up to 610 meters); in 1980 the Sea-Dweller 4000 (water resistant up to 1,220 meters), and in 1983 the GMT-Master II. The Explorer II likewise appeared during the 1970s; this watch—specially designed for cave explorers—had a twenty-four-hour display. In 1992, a new product name was created with the "Yacht-Master," while the "Air-King" model, an inexpensive Oyster without a date display, is among the models that have been manufactured over the longest period.

Many of these models are still in the product range down to the present day. They are consistently being refined and improved, yet often the layperson can hardly distinguish them from the original models. The "Milgauss," for which the first production cycle ended in 1988, has also been available again in a redesigned version since 2007.

ADVERTISING MEDIA

Rolex understood early on how to win over prominent customers—be they athletes, scientists, artists, adventurers, or extreme mountaineers—for the brand. The most successful motor sportsman of the 1930s, Mercedes-Benz in-house racing driver Rudolf Caracciola, arrived at the starting line wearing a Rolex chronograph, and the "fastest man in the world," speed record holder Sir Malcolm Campbell, did likewise, along with world-class golfers Arnold Palmer, Jack Nicklaus, and Gary Player. In collectors' circles, the chronographs worn by ski racer Jean-Claude Killy and actor Paul Newman are nowadays associated with the names of their wearers.

When Sir Edmund Hillary and Tenzing Norgay first reached the summit of 8,848-meter-high Mount Everest in 1953, a Rolex came along with them. The same was true when the "Trieste" bathyscaphe made its dive in 1960, when Jacques Piccard penetrated the Pacific to a depth of 10,910 meters.

A basic automatic movement was the movement most frequently used for manufacturing the multivariant watches of the sports watch family; these models would be modified appropriately for their intended use, such as providing them with magnetic field resistance, a second time zone, a twenty-four-hour display, and so forth. This development began in 1950 with caliber 1030, which had a diameter of 28.5 mm and a height of 5.85 mm. The oscillation frequency was a moderate 18,000 vibrations per hour (vph). The movement had twenty-five jewels and a screw

Lower pictures: The "Kew A" caliber was specially developed in the late 1940s for chronometer competitions.

balance with timing washer, a Breguet overcoil hairspring, and chronometer regulation. The caliber 1530 followed in 1957, with the same diameter but with twenty-six jewels. In 1963, for calibers 1520 and 1580—derived from the former movement—the half-oscillation rate was increased to 19,800 vph. A screw balance with two regulator screws ("Micro-Stella") was used in caliber 1555, which was also derived from caliber 1530.

A completely new movement was presented in 1977 under the caliber number 3035 (28.5 mm diameter, 6.35 mm height).

It had twenty-seven jewels, and the number of oscillations was now 28,800 vph. The caliber 3135 represented the last stage of upgrading this series. It was presented in 1990 and once again featured improvements in terms of numerous details. This conservative model policy is reflected in the quality: the sophisticated technology itself guarantees the unconditional reliability and excellent rate values of Rolex watches.

A WORLD UNTO ITSELF

Chronographs have always played a special role in Rolex sports watches. Not only are these models coveted and sought after as new watches (especially the artificially rare versions in stainless-steel cases); historical models also achieve top prices at auction—up to twenty times their original retail price.

The early Rolex chronographs with single-pusher controls (dating from around 1930) were equipped with the thirteen-line VZ ("Valjoux Zähler," or "Valjoux chronograph") or the 10.5-line 69, both column wheel calibers made by Valjoux. In keeping with the taste of the times, these were delicate watches with a diameter between 28 and 32 mm. It was not until the VZ caliber was replaced by the Valjoux 22 (fourteen lines, corresponding to a 32 mm diameter of the movement) that the watches became larger, and square cases were used besides the traditional round case

Attached to the hull of Jacques Piccard's bathyscaphe *Trieste*, a special Oyster with a reinforced case and domed crystal reached the deepest point of the Pacific Ocean in the Mariana Trench, 10,910 meters below sea level.

In 2003, Rolex presented an anniversary Submariner to mark the fiftieth birthday of this successful model. *At center,* the Rolex Oyster Perpetual Submariner, the original model from 1953. *On the right,* self-winding automatic caliber 3135, which has long been used in most of the Oyster models for men.

Record holder: In 2017, a Daytona with a "Paul Newman" dial that had been owned by the eponymous actor sold for $17.7 million at auction, thus becoming the most expensive stainless-steel wristwatch in the world.

Left: In 2000, the Daytona received its own Rolex caliber with the designation 4130.

for chronographs. At 44 mm, the rare split-seconds chronograph with the Valjoux 55 vbr caliber was imposing, even by today's standards.

After the Valjoux 22 and 23, the Valjoux 72 followed around 1948; this was the caliber that was to power most of the Rolex chronographs and their derivatives right up into the 1980s, from the calendar watch, the "Cosmograph" (in the beginning without the Daytona suffix), to the last manual-winding Daytona (References 6265 and 6263). The watch with caliber 727 was even offered in a gold version, both in 14- and 18-karat gold, as a certified chronometer for the first time.

As the Valjoux 72 C, with month, day of the week, and date display, the movement was also used in the so-called "pre-Daytona." The Reference 81806, with moon phase, which was on the market at the beginning of the 1950s, was made in only very small numbers and is exorbitantly expensive today, although the Valjoux 88 movement used in this watch was also available in other brands.

Rolex presented the first self-winding automatic chronograph in its history in 1988, with a case diameter of 40 mm and a thirteen-line automatic movement based on the Zenith 400 ("El Primero"). In contrast to the Zenith basic movement, the balance wheel in the Rolex caliber 4030 oscillated at a reduced frequency (28,800 instead of 36,000 semi-oscillations or vibrations per hour) to improve oil retention and improve reliability; for all that, all the models were now certified as chronometers.

On the first models, the tachymetric scale engraved on the bezel ranged from 50 to 200 km/h, and later from 60 to 400 km/h.

In 2000, with its chronograph caliber 4130, Rolex presented a completely in-house development for the first time. This

The so-called pre-Daytonas: manual-winding chronograph from around 1935; the "Jean-Claude Killy" calendar chronograph (1954), named after the French skiing ace; and *on the right*, the underappreciated James Bond Cosmograph (Reference 6238) from 1966

chronograph, controlled by a column wheel, is made flatter than its predecessor and also has a larger power reserve (seventy-two hours). The modern design features a balance bridge and a vertical friction clutch, which keeps the starting seconds hand from jumping and also makes it possible to operate the chronograph continuously wear-free.

STAMP COLLECTORS

Similar to the ways of philatelists, collecting Rolex "tool watches" is often a matter of small things, tiny special features, and rare (limited) versions. Marginalia are sometimes the decisive factor determining significant price differences of several thousand dollars.

A manual-winding Daytona (available in the References 6265, 6264, 6263, 6262, 6241, 6240, and 6239) with a dial with a strangely empty appearance—reverently referred to by collectors as the "Paul Newman" because he had once won this watch as a prize in a car race and was subsequently displayed wearing it in a photograph on the cover of an Italian magazine—costs well over twice as much as its technically absolutely identical fellow models. Among the pre-Daytonas, it is the Reference 6236, with a full calendar and steel bracelet—referred to by insiders as the "Jean-Claude Killy"—that makes collectors' hearts beat faster and attains up to $88,000 at auctions.

In the case of the Submariner, early models with no crown guards are increasingly in demand, since it became famous when Sean Connery wore it on his wrist in *Dr. No*. If the lettering in "Submariner" is red, this likewise means a significant surcharge. If the watch in question even involves a "Sea-Dweller" with the imprint of "Comex," the legendary company that pioneered industrial deep-sea diving with its oxygen and helium mixture, and if the watch is appropriately documented, the price can break through the $100,000 barrier. Collectors will even pay attention to small details—such as whether the depth indicator on the Submariner is first given in feet or meters—and distinguish between "feet first" and "meter first" watches. A Submariner with a blue-gray bezel instead of the black one, a "Paul Newman" with a red dial, referred to in insider circles as the "Spirit of Japan," an early Turn-O-Graph, or an original Milgauss—all these (always assuming they are authentic) are models that fetch extremely high prices, and the trend continues to rise.

One watch was denied celebrity status even though it played a major role in a James Bond film. This involves the Reference 6238, a chronograph with a steel bracelet and silver dial that Bond wore in the film *On Her Majesty's Secret Service*. It may be the case that this beautiful chronograph didn't become a cult watch because it wasn't Sean Connery playing the role of 007, but rather the relatively unknown Australian actor George Lazenby.

Dr. Harry Niemann

After the Submariner and the Sea-Dweller, the GMT-Master II was also caught up in the boom of "tool watches" with rotating bezels. In the meantime, there are now waiting lists for all of the new models mentioned, and classic models achieve top prices on the preowned market.

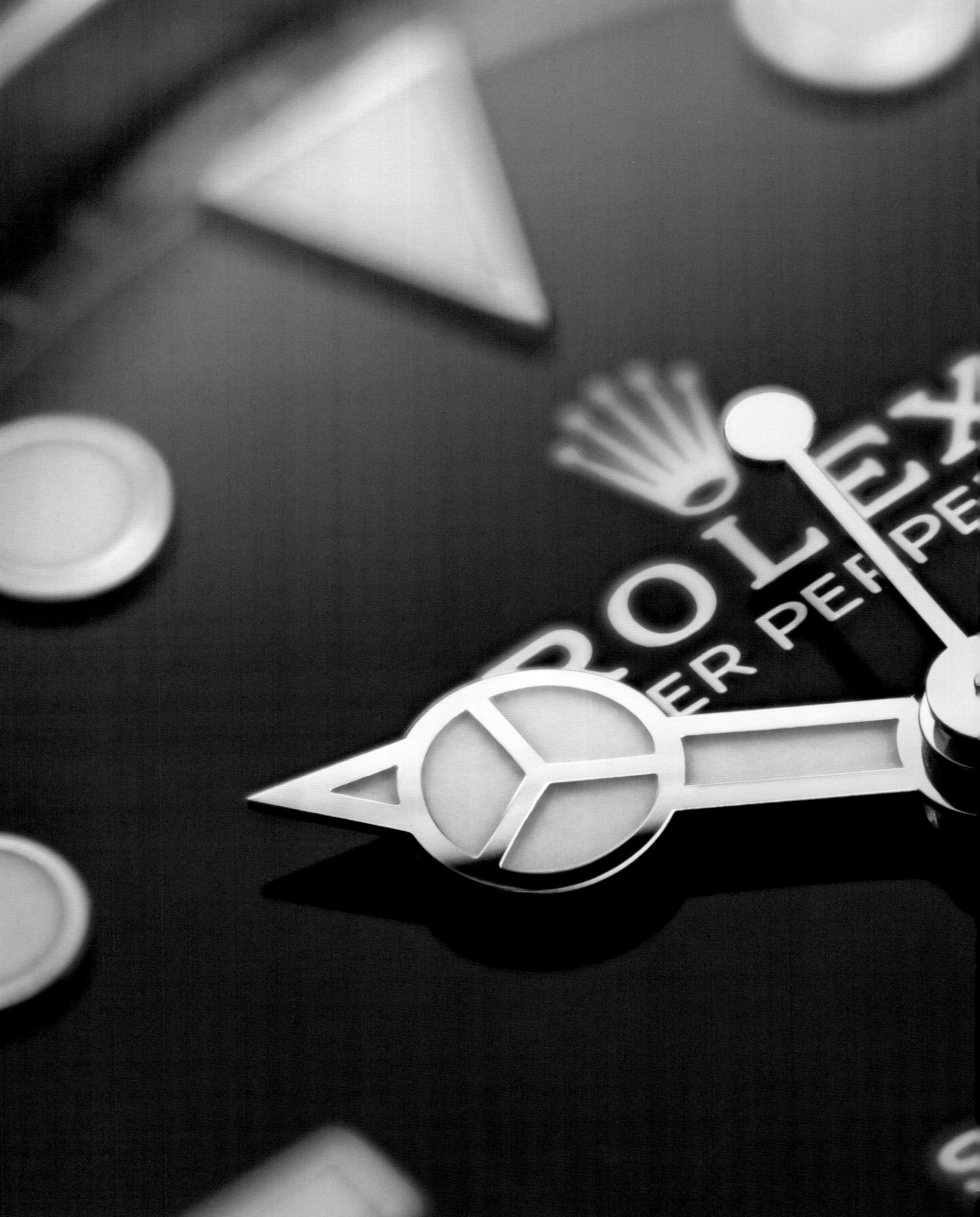

III.
Key to Success

Rolex Tool Watches

The reputation of the Rolex brand is now more than ever based on the professional timepieces called "tool watches," which can actually be used as "tools" due to their robustness, water resistance, and precision of movement.

A Facelift for the Icon: The New Submariner

New proportions and new color combinations: Rolex has modernized the Submariner and the Submariner Date, both in terms of their looks and their interiors. A new movement is being put into action in this legendary diving watch. But this is not the only highlight of the Rolex innovations for 2020.

The dial of the new Submariner is as luminous as ever thanks to Chromalight, the luminous material displayed on the hour markers and hands.

Diving in style? For decades there has been a preferred companion for seasoned divers to wear on their wrists—the Rolex Submariner. Functional and timeless, this icon is the epitome of the diving watch. Its history predestines it for this role: In 1953, Rolex presented the first Submariner, at that time water resistant to 100 meters and with a 38 mm case. The diameter was increased to 40 mm in 1959, and it has remained true to this dimension for a long time.

The Submariner has hardly changed over the decades in terms of appearance, either: the hermetically sealed Oyster case, the unidirectional-rotating diving bezel, and the easy-to-read dial. All are details that determine the appearance of this model, then as now. Rolex never even attempted any big experiments when it came to color: The Submariner palette comprises black, blue, and green.

This consistency has contributed to the everlasting Rolex style. But things have never stood still—on the contrary. Evolution takes place on a small scale, sometimes in secret. Thus, the watch movements have been continually optimized, and from this time on it is the latest generation of in-house calibers from the house of Rolex that is at work in all the Submariner models.

The watch functions have also been under continual development, for ever-greater functionality and wearing comfort. And in terms of materials, they have been applying the latest and greatest over the decades, including a hard ceramic insert for the bezel and a sapphire crystal over the dial.

THE NEXT CHAPTER

Now the icon has taken the next step: the latest generation of Submariner and Submariner Date watches comes in new dimensions. The case has been enlarged by 1 mm to a diameter of 41 mm, and the proportions of the bracelet and dial have been adjusted accordingly. Slimmer crown guards, as well as narrower lugs, contribute to the more modern overall impression.

On the new Submariner Reference 124060, these innovations are hardly noticeable at first glance, especially since the dial and Cerachrom rotating bezel have been kept in black, as on the predecessor watch.

At 1 mm larger in diameter, cleverly concealed by the narrow lugs, and a slimmer flank guard for the screw-down crown, the new Submariner remains true to its lineage.

Oyster Perpetual Submariner

Reference: 124060

Movement: Self-winding, Rolex caliber 3230; 29.1 mm diameter; 31 jewels; 28,800 vph; Chronergy escapement, Parachrom hairspring, Paraflex shock absorber, balance wheel with Microstella regulating nuts; power reserve 70 hours

Functions: Hours, minutes, center seconds

Case: Stainless steel; 41 mm diameter; bezel with ceramic insert, unidirectionally rotatable, with 60 index scale; sapphire crystal; screw-down crown; water resistant to 30 bar

Bracelet: Oyster stainless steel, folding clasp with extension link

Price: $8,076

The most important innovation involves converting the Submariner technology for the new in-house caliber 3230 (version without date display) with Chronergy escapement.

If you put the predecessor model and the innovation next to each other, the changes can be seen almost exclusively in the slimmer crown guards.

When looking at the dial, there is one detail that makes it possible to distinguish between old and new: The dial is signed at the very bottom, below the "6," with the inscription "Swiss made," which the maker has added to the new Submariner along with its trademark, the crown.

This logo is a reference to the completely new inner workings of the Submariner: The model is now equipped with the latest generation of self-winding movement, the newly developed caliber 3230, which is based on the caliber family 3200, launched in 2016. It replaces the previously used caliber 3130 and now offers all the modern details of Rolex watch movements. These include the Parachrom hairspring and the Paraflex shock absorber, as well as a power reserve of an impressive seventy hours. And the price? For the new Submariner, this is $8,076.

OTHER VERSIONS OF THE SUBMARINER DATE

Also new is the Submariner Date; seven different versions of this watch are being presented in stainless steel, gold, and two tone. These models have also grown from 40 to 41 mm—combined with the corresponding new proportions for the case, bracelet, and dial. They are equipped with the already well-known Rolex self-winding caliber 3235, also a representative of the new generation of movements. This caliber offers the corresponding features, such as a seventy-hour power reserve, Parachrom hairspring, and Chronergy escapement.

There are no surprises in terms of the color palette, consisting of black, blue, and green, but in new combinations. For example, the white gold Submariner Date combines a black dial with a blue bezel, and the yellow gold Submariner Date innovations are available in blue or black, as are the two-tone models in stainless steel and yellow gold. Most of the attention will probably be drawn to the green and black innovations: The stainless-steel Submariner Date combines a black dial with a green Cerachrom bezel. This color combination was already part of the collection from 2003 to 2010; at that time it was known by the nickname "Kermit." Now it is back again and is available at a price of $9,560.

Iris Wimmer-Olbort

Color comes into play in the Submariner Date, with colored Cerachrom bezels in green or black for the stainless-steel version, or in blue for the "Rolesor" versions in stainless steel with elements of yellow or white gold.

The History of the Diving Watch

A new model made its debut in the Rolex Oyster family in 1953; it was specially designed for the requirements of professional divers. The most important innovation for this watch, which is water resistant up to 100 meters, was a rotating ring around the outer edge of the crystal that could be used to mark and take note of diving and decompression times. The Rolex Submariner was born, and with it a new category of watches.

Water-resistant watches had been a specialty of the house since the mid-1920s. It was not without reason that Rolex founder Hans Wilsdorf named its robust stainless-steel case with the elongated body and round glass holder ring after the oyster, the symbol of perfect reserve and closure.

With a large screw-down back, a screw-down crystal holder circlet, and a crown that likewise was screw down, the "Oyster" set the standard for all future water-resistant watches. "The Wonder Watch That Defies the Elements" ran the headline of the *London Daily Mail* on November 24, 1927, next to the portrait of English Channel swimmer Mercedes Gleitze and a selection of four Rolex "Oyster" watch models. Wilsdorf had been working toward this moment for a long time, and when the daring young woman made this spectacular appearance along with the Oyster, he made his move: for 40,000 Swiss francs he secured the front page of the daily newspaper, which is published in the millions of copies, and made his brand famous overnight.

When they had perfected the automatic, self-winding watch movement by using an oscillating-weight rotor, Rolex engineers took another step in the direction of creating a robust and problem-free wristwatch at the beginning of the 1930s. The self-winding watch movement, which performed its work without external intervention and supplied itself with energy via the motion of the wearer's forearm, was an important precondition for ensuring the long-term quality of the watch. The screw threads of the crown and the case tube were spared and were largely eliminated as the last remaining sources of any possible malfunctions.

The rather small Oyster models (mostly less than 33 mm in diameter), also known as the "Bubble Backs" because of their domed case back, enjoyed great commercial success in the 1930s and 1940s. The "Oyster Perpetual" self-winding versions quickly outstripped the manual-winding versions, and Rolex became synonymous with water-resistant automatic watches. They also became famous for their precision according to chronometer standards.

From this standpoint, even the most-elegant and most-delicate watches made by Rolex were the true sports watches of their times, although today we have a different image of sports watches. But it was precisely this image that was yet to be shaped—by Rolex.

There was no such thing yet as an over-the-counter diving watch for professional use at the beginning of the 1950s. Professional divers used mostly military watches in their work—bulky devices with thick crystals and pocket watch movements, or even just timers. These timepieces had little in common with "wearable" wristwatches.

PROFESSIONAL FUNCTIONS

It was in the early 1950s when Rolex first equipped an Oyster Perpetual with a rotating bezel and named this watch, with Reference 6202, the "Turn-O-Graph." It was the first model in the family of "professional watches," because with its clear functionality, it was aimed at users who required a means for determining periods of time, but for whom a chronograph—which Rolex did indeed make, even if not in a water-resistant version—would be too expensive or too delicate because of its technology.

The rotating bezel on the Turn-O-Graph had a sixty-unit scale and a triangular zero marker coated in luminous material that could be adjusted to the tip of the minute hand and thus helped in measuring shorter time intervals. The dial ring did not have any ratchets, nor was there any other way to prevent the wearer from rotating it unintentionally—which, by the way, was still possible to do in both directions at that time. The screw-down crown already featured the "Twinlock" seal, which held tight to a depth of 50 meters even if it wasn't screwed down properly.

The style-defining Submariner: *on the far left,* one of the first Oyster Perpetual Reference 6204 watches with "Submariner" inscription on the dial, but still without crown flank protection on the case (1953). *In the center,* the Submariner Reference 5512 from 1959, as it was manufactured almost unaltered over several decades; *on the far right,* a Submariner Date from 1983.

Three generations of Submariners: *above,* an early chronometer version from 1955, including two Submariner Date watches from 1969 (gold) and 1979 (stainless steel)

The twenty-fifth anniversary of Professor Piccard's record dive in the *Trieste* was properly celebrated by Rolex in 1985.

The successor model, with Reference 6204, was explicitly intended for underwater use. This model was presented a little later on and then also bore the designation "Submariner" on the dial—initially still without a depth indicator. In 1953, the year it was born, the Submariner still had the normal Oyster case with a screw-down Twinlock crown; however, it was approved by Rolex for a diving depth to 100 meters. The generous coating of the hour markers and pointers with luminous material ensured good readability even in adverse lighting conditions.

Thus, for the first time, the Geneva-based brand had given professional divers the opportunity to purchase from specialist watch dealers a professional timepiece tailored to their needs. With the help of the rotatable bezel, it was possible to mark diving and decompression times, take note of them, and read them off at a glance. This important "diver's bezel" function, meeting numerous standards and military regulations, was later established as an essential part of a diving watch.

SHORT-TERM EVOLUTION IN A FEW STEPS

The Submariner was given its characteristic face with the "Mercedes star" on the hour hand with the introduction of the Reference 6536 during the mid-1950s. This model already had fine scaling for the first fifteen minutes on the rotatable bezel, something that individual national diving-watch standards would prescribe later on.

Modern-day Submariner models without (*far left*) and with date display in various color combinations

Rolex has remained true to the maritime theme in its advertising motif for all these years, even though the Submariner has long since enjoyed universal recognition as an all-around sports watch.

By 1965, people had already realized that the Submariner could do more than just "keep its mouth shut."

The Reference 5512, in production from 1959 onward, was the first to have a case with flank guards for the screw-down crown. The case diameter had also been increased from 37 to 40 mm, and besides this, Rolex had meanwhile guaranteed water resistance down to 200 meters. The caliber of the 1500 family, which had been revised in many respects, replaced the old model with 9¾ lines.

The newly designed rotatable bezel, which was easier to grasp, had a system called "Auto Lock" that prevented the wearer from rotating it unintentionally: To adjust the dial ring, you had to press it against the case; after it was released, it snapped back into place.

This model finally advanced to become standard equipment for professional divers both in civilian and military environments and remained in the range with only minor changes until 1991. For example, the Submariner Date (Reference 1680) was introduced in 1969, and a Submariner in a gold case was offered for the first time. In 1979, the Plexiglas was exchanged for a sapphire crystal, increasing the water resistance depth to 300 meters.

At the end of the 1980s, the Reference 14060 finally became the heir to the Submariner—with a unidirectional rotatable bezel, a screw-down Triplock crown, and the (provisionally) last evolutionary stage of the self-winding caliber 3000.

FORMATIVE INFLUENCE

The availability of this robust professional watch at normal watch dealers ensured that the Rolex Submariner was already widely distributed as early as the 1960s. After the sports and hobby divers, sailors, and motorboat captains discovered the qualities of the diving watch, it did not take long before other recreational athletes became aware of this robust, low-maintenance, and reliable timepiece. Film and television heroes such as Sean Connery, Roger Moore, and Paul Newman were happy to adorn themselves with these "professional watches" and made their own contributions to the Submariner's popularity. In collectors' circles, the early models (without flank guards for the crown) are sometimes referred to as "James Bond" Submariners.

The Rolex Submariner fit into the active lifestyle of the 1960s and 1970s in a way that no other watch did, and numerous fellow models, from the Yacht-Master to the GMT-Master to the Sea-Dweller, struck the same chord. In advertisements, Rolex encouraged private and civilian customers to buy what are basically highly specialized watches, using the motto "What's right for the professionals can only be a bargain for you."

Speaking of being a bargain: since Rolex professional watches were by no means any kind of special-bargain offer—although the level of the prices during the 1960s brings tears to our eyes today—owning a Submariner had not only a sportswear but also a luxurious component. The mystique of the "luxury sports watch" persists to this day, although there have long been more-luxurious, more-exclusive, and incomparably more-expensive sports watches available to buy.

The logical evolution of the watch's appearance, the unsurpassed readability of the displays, and the convincing ways in which the functions work all ensure the Submariner the status of an archetype among diving watches.

In fact, it really isn't possible to improve this watch—at most, it can only be complemented. For example, by making versions in gold, or steel and gold ("Rolesor"), or in a genuine fiftieth-anniversary model—with a green bezel, another real sensation in 2003!

Peter Braun

The King Is Back

Under the crown logo in yellow and the brand name in green, the inscription "Air-King" makes its appearance once again today, just as it did when especially created for this model in the 1950s. The Explorer has also been redesigned—with completely luminescent indices on the dial instead of a mixture of appliqués and lacquered indices as featured on the Air-King.

Above: Since 2016, the new Explorer has made quite a straightforward appearance, without any highlights on the dial, but on the other hand, it consistently has featured luminescent markers.

Opposite page: Three polished hour markers and a wedge index at the "12" are the distinguishing features of the Air-King–as well as, of course, the yellow crown with the brand name inscribed in green.

The Air-King stands for the special relationship that linked Rolex to the world of aviation during its golden age in the 1930s. This period was marked by rapid development in aircraft performance, leading to the first record-breaking and long-distance flights. The new Air-King also pays homage to the pioneers of aviation and to the role that the Rolex Oyster played in the heyday of aviation—as a reliable chronometer that fearless pilots wore on their wrists.

Despite the fact that its theater of operations was in the "wild blue yonder," the Air-King, like all Oysters, has always been water resistant and equipped with a screw-down crown. Besides this, it was always made with a soft-iron inner cage to protect its movement from the strong magnetic fields generated by the ignition coils for large-volume combustion engines in the cockpit, depending on where they were mounted in the engine. This also applies to the self-winding automatic movement of caliber 3131, which is used in the new Air-King.

When redesigning the dial, the three hour markers for 3, 6, and 9 were designed as polished metal appliqués. In good lighting conditions, these markers can hardly be distinguished from the minute numerals printed using a luminous material that complete the hour circle.

In a dimly lit environment the three markers recede into the background of the matte black dial because of the dark reflection, while the minute numerals emit a bright luminescence. But to make the new dial of the Explorer, a close relation of the Air-King, Rolex has once again abandoned this mix of materials. On this dial the three hour markers, as well as the index markers, are now coated with luminescent Ceralight material in blue, making reading considerably easier during night and day.

The two sports tool watches, the Explorer and the Air-King, have also been tested and certified as "superlative chronometers" in accordance with new precision standards. Ultimately all the inscriptions on any Rolex watch dial, from "Oyster" to "Perpetual" and all the way to "Superlative Chronometer Officially Certified," are guarantees of performance. These are the chapter headings for the history of a mythos of the art of watchmaking. With a recommended retail price of $6,432, the Air-King is a real bargain, because the upgraded Explorer model costs $6,542 without the magnetic field shield.

Peter Braun

Oyster Perpetual Air-King

Reference: 116900

Movement: Self-winding, Rolex caliber 3131; 28.5 mm diameter; 31 jewels; 28,800 vph; Parachrom hairspring, Glucydur balance wheel with Microstella regulating nuts; magnetic field resistance due to the soft-iron inner cage; power reserve 48 hours; certified chronometer (COSC and Rolex)

Functions: Hours, minutes, center seconds

Case: Stainless steel; 40 mm diameter, 11.2 mm thickness; sapphire crystal; screw-down crown; water resistant to 10 bar

Bracelet: Oyster stainless steel, folding clasp, with safety catch and extension link

Price: $6,432

Pioneering Achievement

For the fiftieth anniversary of the Sea-Dweller in 2017, Rolex presented a new version of this legendary deep-sea diving watch, which has effectively been part of the standard equipment for professionals since 1967. Equipped with a latest-generation self-winding automatic movement, with the case enlarged to 43 mm and an extra-thick sapphire crystal fitted with a date magnifier, the new Sea-Dweller has started on its way into the next fifty years.

There came a time when the water resistance of a Submariner, guaranteed to 200 meters, was suddenly no longer sufficient for professional divers, because their professional lives were taking them to ever-greater depths. Breathing air mixtures with high nitrogen and helium content mitigated the risk of oxygen poisoning, posing a threat when diving to depths of 60 meters, and new findings regarding decompression times made in the 1960s opened up a new world for people in the depths of the oceans. Divers were spending more and more time underwater in diving bells or deep-sea stations to explore the underwater world, or to dig up natural resources.

DWELLERS OF THE SEAS

A conventional diving watch such as the Submariner not only offered an insufficient level of water resistance; it came up against a physical limit, especially in the decompression chamber. During the long times spent in the overpressure of the diving station and decompression chamber, the very small helium molecules in the breathing air penetrated past the watch seals into the interior of the watchcase. If the ambient pressure then decreases again, the gas cannot escape quickly enough and literally bursts the watch crystal out of its setting.

The new generation of diving watch that Rolex introduced in 1967 had an automatic one-way valve in the watchcase that allows pressure to be equalized in one direction while ensuring a 100 percent tight seal in the other direction. The so-called helium valve became the trademark of the Sea-Dweller.

A NEW INTERPRETATION OF A LEGEND

When it was last brought up to date in 2014, the Sea-Dweller had already been adapted to the new standards of Rolex diving watches. At the same time, the anniversary model Sea-Dweller 2017 appeared in a completely newly designed version: the case, which was still water resistant to a depth of 1,220 meters, was enlarged to 43 mm in diameter, making it possible to expand the dial display accordingly in the interest of even-better readability. The entire watch has been newly dimensioned,

ROLEX
SEA-DWELLER
4000ft = 1220m
SUPERLATIVE CHRONOMETER
OFFICIALLY CERTIFIED
SWISS MADE

Since 2017, the Sea-Dweller has also featured a magnifier polished into the crystal over the date window.

while Rolex lent an eye to aesthetic and functional balance, from the width of the rotating bezel, the bracelet, and clasp, to the thickness of the cover crystal. For the first time in the history of the model, the crystal was fitted out with a magnifier polished into the crystal over the date window and thus visually matched the Submariner Date.

These features correspond to the highest technical standard for Rolex diving watches to the present day. The case and bracelet are made of 904L stainless steel, distinguished by its extreme corrosion resistance. The bezel, which can be rotated only counterclockwise, has a Cerachrom number disc made of extremely scratch-resistant and highly durable ceramic. The hour markers and hands are coated with long-lasting Chromalight luminescent material, ensuring optimum readability even in the dark. The Oysterlock safety folding clasp prevents the wearer from opening the bracelet accidentally and, in addition to the finely stepped Glidelock fine-adjustment system, also has a Fliplock extension link that can be folded out. Altogether, these features make it possible to extend the bracelet by up to 46 mm without using a tool, so that the watch can be comfortably worn on top of a 7 mm thick diving wetsuit. The screw-down Triplock winding crown is equipped with a triple sealing system, and of course the new Sea-Dweller also boasts the component that has given it its reputation: the helium valve.

LATEST-GENERATION WATCH MOVEMENT

The new Sea-Dweller was the first "tool watch" to benefit from the new generation of movements. After the elegant sports Day-Date 40 and Datejust 41 models, in 2017 the first diving watch was also equipped with a movement from the caliber 3200 family. A 15 percent higher level of efficiency for the escapement, 50 percent more power reserve, 90 percent newly developed parts, and a whopping 100 percent higher level of precision than the COSC (Contrôle officiel suisse des chronomètres) requires made the Sea-Dweller with Reference 126600 a veritable chronometer of the superlative.

The movement diameter has been increased slightly compared to the previous calibers of the 3100 family (from 28.5 to 29.1 mm), but it was possible to decrease the overall height from 8 to just over 5 mm. There are fourteen patents protecting the innovative design details; in total, these innovations not only

Sea-Dweller

Oyster Perpetual Sea-Dweller

Reference: 126600

Movement: Self-winding, Rolex caliber 3235; 29.1 mm diameter; 31 jewels; 28,800 vph; Parachrom hairspring, Paraflex shock absorber, Chronergy escapement, Glucydur balance wheel with Microstella regulating nuts; power reserve 70 hours; certified chronometer (COSC)

Functions: Hours, minutes, center seconds; date

Case: Stainless steel; 43 mm diameter, 13.8 mm thickness; bezel with ceramic insert, unidirectionally rotatable, with 60 index scale; sapphire crystal; screw-down crown; helium valve; water resistant to 122 bar

Bracelet: Oyster stainless steel, folding clasp with Glidelock fine-adjustment mechanism and Fliplock extension link

Price: $11,700

Luminous hands and hour markers make the Sea-Dweller a reliable diving partner at great depths.

The solid Oyster link bracelet has an Oysterlock clasp featuring the Glidelock (fine adjustment) and Fliplock (safety extension link).

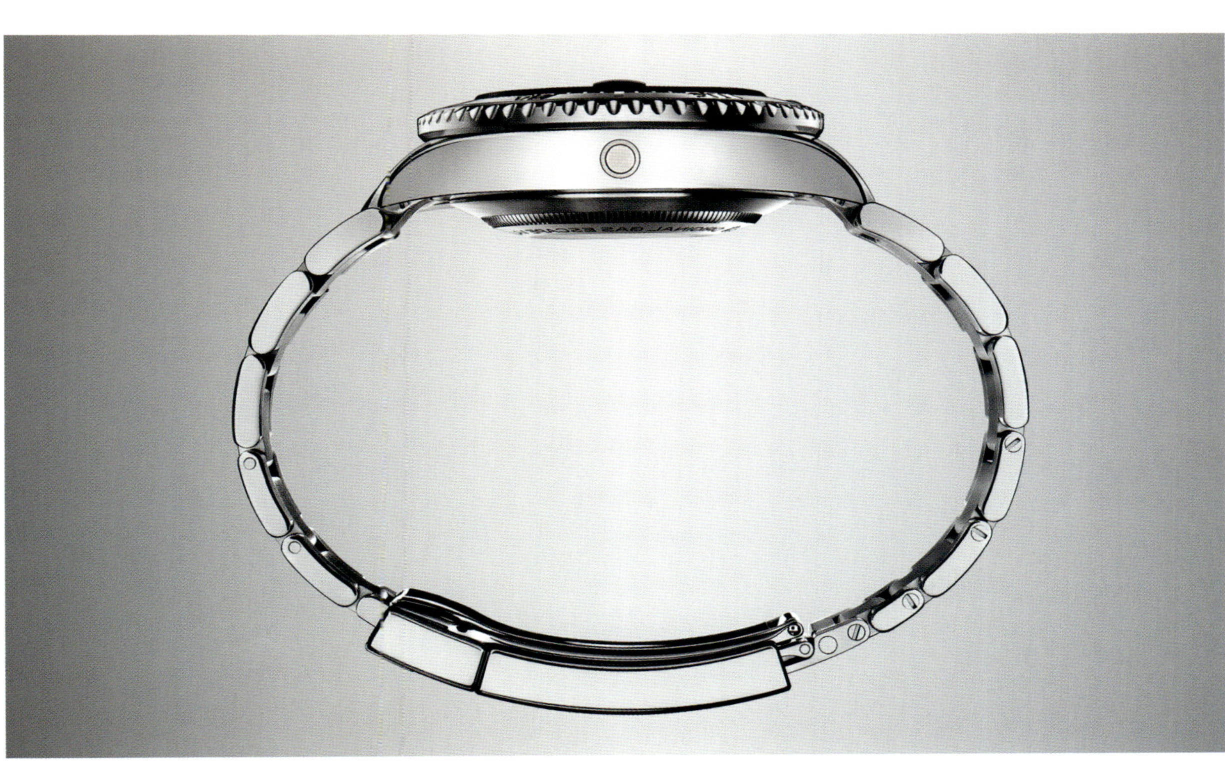

The automatic helium valve is embedded flush into the left flank of the Sea-Dweller watchcase.

have increased the power reserve to a full seventy hours but also have improved rate precision in the various positions and at all temperatures, thanks to the more efficient escapement and the friction-optimized wheel train bridge. The core of the new design is the escapement patented under the name Chronergy, which is in fact based on the principle of the Swiss lever escapement, but thanks to its new geometry and systematic lightweight construction, it is 15 percent more efficient. The anchor and escapement wheel, the delicate key components of the Chronergy escapement, are manufactured from a nickel-phosphorus alloy by using what is called in German the LIGA process (*Lithographie*, *Galvanoformung*, *Abformung*: lithography, electroplating, and molding; i.e., electrogalvanic molding) and are therefore insensitive to magnetic fields.

The caliber 3235 balance wheel features a blue Parachrom hairspring manufactured at Rolex from a patented niobium-zirconium alloy. The nonmagnetic hairspring is temperature stable and is made to be up to ten times less sensitive to shocks than a conventional hairspring. The oscillating system is held in place by a Paraflex shock absorber in a traversing balance bridge and features an optimized height adjustment system.

With this new generation of watch movements, Rolex set a new standard in the field of precision that goes beyond the criteria of the independent Official Swiss Chronometer Testing Institute, or Contrôle officiel suisse des chronomètres. The brand has developed new high-tech processes and devices so that it is able to ensure the precision of its "superlative chronometers" with tolerances that are twice as strict as those of the COSC (+2/–2 seconds per day instead of +6/–4 seconds) and to test them under conditions that simulate a real-life situation that corresponds more closely to the everyday life of the wearer. Rolex has developed a specific test protocol on the basis of large-scale statistical studies to determine the real conditions that occur when wearing a wristwatch. Ever since then, these exclusive tests have been applied to supplement the official certification by the COSC, to which all Rolex movements are still systematically subjected. After the movement has been encased, the tests are extended to the entire fully assembled watch.

For all the love of technical progress, this impressive anniversary model has succeeded in preserving the charm of the pioneering era—with the "Sea-Dweller" inscription on the black dial in red, as it was on the first historical versions.♛

Peter Braun

The caliber 3235 belongs to the latest generation of movements with Chronergy escapement, blue Parachrom hairspring, and Paraflex shock absorber.

In terms of price and performance, the Sea-Dweller ranks between the Deepsea (*above*) and the Submariner (*below*).

THE HELIUM VALVE

Professional saturation divers experience a troublesome phenomenon during the decompression phase after deep dives that also affects their watches. To slowly readapt to normal atmospheric pressure at the surface, divers must spend extended periods in a decompression chamber, where they inhale the gas mixture called heliox, consisting of about 95 percent helium. The extremely small helium molecules are able to penetrate past the case seals into the interior of the watch, causing the pressure to slowly rise to the ambient level. During the subsequent decompression of the chamber, the gas isn't able to escape from the waterproof watchcase fast enough, so that the watch crystal is often literally blown off like a champagne cork by the resulting overpressure.

Rolex engineers invented and patented a small valve that ensures a controlled escape of the trapped helium during the phases of decompression, without compromising the watch's absolute waterproof capacity. The Sea-Dweller was the first diver's watch to benefit from such an automatic helium valve, perfectly meeting the needs of deep-sea divers.

Externally, the Sea-Dweller resembled the Submariner: a stainless-steel case 40 mm in diameter, black dial with luminescent indices, and rotating bezel with sixty minute graduations. The case had a slightly thicker wall than that of the Submariner, and the Sea-Dweller also had a date display—an important feature for saturation divers, who perform their work for several weeks at a time in isolated hyperbaric habitats. Rolex also equipped the watch with another innovation: a bracelet with flip-out extension links ("Fliplock") that allow divers to wear their watches either under or over the thick diving suit that protects them from the cold at these extreme depths.

ROLEX
GMT-MASTER II
SUPERLATIVE CHRONOMETER
28
SWISS MADE

Pepsi for the People

At Baselworld 2018, Rolex presented—finally!—a stainless-steel version of the GMT-Master II with a red-blue rotating bezel. Previously the coveted "Pepsi" version was available only in a prohibitively expensive white gold design. There is still a small downer, however . . .

Diehard and loyal friends of the brand are so easy to satisfy. It doesn't always have to be a new watch movement, a new model, or a new collection. Sometimes a new color of the dial is enough to put them in raptures. Or even a red-blue bezel.

JUST FOR MILLIONAIRES AT THE START

Four years ago, when Rolex gave its successful GMT-Master II model a new twenty-four-hour rotating bezel in two-tone Cerachrom, the watch world was stood on its head: blue and red—those were the colors of the original model from 1955, which trades among collectors for expensive prices under the nickname "Pepsi." The actual news, which was much more exciting, faded completely into the background in the face of the praise for the sense of tradition. In fact, Rolex had already gained some experience with the high-tech ceramics used to make other sports models from the Oyster family, but the combination of blue and red was an absolute novelty. Never before had chemists succeeded in creating a richly luminous red ceramic, and just a few weeks earlier another watchmaker had boasted of its "unique" case made of red ceramic.

Oyster Perpetual GMT-Master II

Reference: 126710 BLRO/69200

Movement: Self-winding, Rolex caliber 3285; 29.1 mm diameter; 31 jewels; 28,800 vph; Parachrom hairspring, Paraflex shock absorber, Chronergy escapement, Glucydur balance wheel with Microstella regulating nuts; power reserve 70 hours; certified chronometer (COSC)

Functions: Hours, minutes, central seconds; additional 24-hour display (second time zone); date

Case: Stainless steel; 40 mm diameter, 13 mm thickness; bezel with ceramic insert, bidirectionally rotatable, with 24 index scale; sapphire crystal; screw-down crown; water resistant to 10 bar

Bracelet: Jubilee stainless steel, "Oysterclasp" folding clasp

Price: $9,665

Other versions: In stainless steel/rose gold ($14,800); in rose gold ($38,440)

The GMT-Master II with a red-and-blue rotating bezel in an inexpensive stainless-steel version was the highlight of the 2018 season for Rolex fans.

The special attraction of the two-color Cerachrom ring in the rotating bezel comes from the fact that the two colors remain within their assigned halves of the bezel without any perceptible transition and are absolutely separate. This is possible only because the ceramic insert is initially monochrome, then the color is changed on a molecular basis in a downstream work step—in this case, from red to blue. The markers and the index points are engraved on the surface and then flushed out with platinum, using the PVD process.

Both brand enthusiasts and connoisseurs therefore had good reason to love the new GMT-Master II "Pepsi" (Reference 116719 BLRO). Now, if it just weren't for that hefty price tag of $34,000: the "Pepsi" was initially available only in white gold with an Oyster link bracelet in white gold.

WITH A COMMON TOUCH IN STAINLESS STEEL

In 2018, Rolex finally introduced the long-awaited stainless-steel version of the GMT-Master II with the "Pepsi" bezel made of Cerachrom. Until now, the mother of all GMT watches in stainless steel was available only with a ceramic bezel imbued with black and blue colors. And here it must be stated very clearly: the new version with the red and blue bezel looks a lot better!

Yet, even if diehard Rolex fans see things differently, for Rolex, things are never just about how it looks—not even with the new Reference 126710 BLRO. As a special treat, the crown-regulated mechanism with the incrementally adjustable hour hand has been integrated into a self-winding movement of the latest generation. The Rolex caliber 3285 is therefore equipped with an efficient Chronergy escapement with an anchor and escapement wheel made of a nonmagnetic nickel-phosphorus alloy. The blue Parachrom hairspring likewise cannot be magnetized and plays its part in the excellent timekeeping performance. It is tested in a separate regulation procedure—after obtaining the COSC chronometer certificate—to plus or minus two seconds when the watch movement is engaged. Fourteen patents protect the design

An unusual sight: a delicate Jubilee bracelet on a tool watch with a rotating bezel–something last seen at Rolex fifteen years ago.

Below:
Modern alchemy. Until now, Cerachrom material wasn't available in red. The sharp transition to blue is achieved by chemically imbuing one-half of the bezel with the color.

details, which in their totality also increase the power reserve to seventy hours. Against this background, the increased price of $329 more than that for the old black-blue GMT-Master II is perfectly all right.

JUBILEE BRACELET ON A TOOL WATCH?

What could Rolex have been thinking when they selected the five-row Jubilee design pattern for the new steel "Pepsi"? This is something that otherwise we know of only on the elegant versions of the Oyster. To top it all off, why did Rolex fit it out with an Oysterclasp safety clasp? Perhaps they wanted to still leave those loyal brand enthusiasts who only a few years earlier had spent a fortune on the white gold "Pepsi" another small and unique selling point? This theory would at least be supported by the fact that it is not easy to exchange the Jubilee bracelet with a normal stainless-steel Oyster link bracelet, due to the slightly modified bracelet lugs.

In the eyes of some, this greatest Rolex model innovation of recent years suffers from its slightly rattling link bracelet, which provides little lateral support. But basically, when you are wearing the watch, this kind of feel fits perfectly with the retro charm of the red-and-blue "Pepsi" bezel.

Peter Braun

IV.

Complicated Watches

Concentrating on a few simple and practical functions was once one of the keys to success for Rolex. In the meantime, a few technical special features have been added on that illustrate the innovative strength of Rolex as a luxury watch manufacturer.

ROLEX
-DWELLER
STER PERPETUAL
ERLATIVE CHRONOMETER
OFFICIALLY CERTIFIED

Perfect Timing

Daytona, Florida, is known as the world capital of speed. Car races were held on its fine sandy beach as early as 1903, and whole series of new speed records were set over the years. The modern Daytona International Speedway hosts the famous twenty-four-hour race every year. Anyone who names a watch after Daytona has a clear idea about things.

One of the icons of the Rolex brand is undoubtedly the Cosmograph, introduced in 1963—the advanced version of the classic chronograph as it has existed in the Rolex collection since the 1930s. The totalizers on the new Cosmograph stood out clearly from the dial in their strongly contrasting colors: Black totalizers were combined with a light-colored dial or light-colored totalizers with a black dial. Today this color scheme is called the "panda" or "reverse panda," as American collectors once vividly described the different designs. And thus we find ourselves in the middle of an extraordinary history.

From Daytona to the *Cosmograph Daytona* —a Passion for Speed

The current Rolex Daytona with contrasting black or white dial and scratch-resistant ceramic bezel

The modern chronograph is gaining in profile: in 1963, as the Cosmograph with black dial and white totalizers; in 1964, for the first time with "Daytona" in red lettering; and in 1965, with a black tachymeter bezel on the "Panda" dial.

WHY DAYTONA?

Just a few months after the introduction of the Cosmograph, Rolex was offering new dials, including one with the additional inscription of "Daytona" in red letters in a semicircle above the twelve-hour totalizer at the "6." This inscription, originally limited to models for the US market, was most likely added at the request of Rolex's subsidiary in the United States to underscore the brand's relationship to the Daytona International Speedway in Florida as its official timekeeper. It was gradually introduced on all the dials, so the term "Daytona" soon replaced the official model name, Cosmograph.

A special version of the Daytona later became known as the "Paul Newman dial" because the famous American actor—both a racing driver and a style icon for his time—wore a Cosmograph with this special dial on an everyday basis. Characteristic of this watch was the minute track on the outer rim of the dial, on a band in the same contrasting color as that of the three totalizers, in some cases with red graduations. The totalizers were typically designed with block-shaped indices to make them easier to read. In 2017, such a "Paul Newman" Daytona from the estate of its namesake set an absolute auction world record for a stainless-steel wristwatch of $17.7 million.

ROBUST AND WATER RESISTANT

The first Rolex wristwatch chronograph from 1939 was already equipped with the water-resistant Oyster case developed by Rolex in the mid-1920s. The Cosmograph was also robust and water resistant thanks to the Oyster case, because the case back and winding crown are screwed down to the middle section. Besides this, it was fitted with a solid metal bracelet as a standard feature. At a time when a chronograph with a self-winding mechanism still represented an unmet technical challenge, the Cosmograph featured a manual-winding Valjoux chronograph movement, known for its reliability and precision.

In 1965, the Daytona experienced a further leap in development with a version with water-resistant, screw-down chronograph pushers, making the Oyster concept complete. Moreover, this prevented the wearer from accidentally pushing the pushers. As an indication of the increased water resistance level, the inscription "Cosmograph" was now supplemented with the addition of "Oyster" on the dial. Another innovation was that the tachymeter bezel was now fitted with a black Plexiglas disc with white graduations, which made it easier to read.

With its own-make chronograph caliber 4130 for the Daytona, in 2000 Rolex set a clear signal as a fully integrated luxury watch manufacturer. To this day, the brand and product are closely linked to the Daytona twenty-four-hour race.

During the 1980s, the manual-winding Daytona was having a difficult time against the self-winding automatic chronographs, which had been available since 1969. In 1988, Rolex finally decided to convert the Daytona to self-winding and commissioned Zenith to produce a version of the "El Primero" caliber that was modified in many technical details, ending up with more than 50 percent of the components being replaced by specially developed parts. The caliber 4030 that resulted had, among other things, a Rolex balance wheel with Microstella fine adjustment and a hairspring with a Breguet overcoil; besides this, the vibrations per hour had been reduced from a wear-inducing 36,000 vph to 28,800 vph. Thanks to meticulous quality control during assembly and systematic tests, the Daytona self-winding movement was certified as an official Swiss chronometer. As a result, all versions of the new model were subsequently inscribed with "Oyster Perpetual Cosmograph Daytona" on the dial, as well as with "Superlative Chronometer Officially Certificated."

The reworked design of the self-winding Daytona became the basis for the appearance of today's Cosmograph. Its Oyster case has been enlarged from 36 to 40 mm in diameter and was given a crown flank guard. The metal tachymeter bezel was made wider and given an engraved graduated 400 unit scale. New hands, new hour markers, new totalizers with borders: the dial became more modern but still retained its incomparable style and its signature in red letters: Daytona.

The incredible success of this new model was undoubtedly due to the mechanical renaissance in the early 1990s—a phenomenon that the Daytona itself was very likely instrumental in creating. And just as happened in 1988, when rumors about a self-winding version fueled the demand for the manual-winding Cosmographs with Valjoux technology, toward the end of the millennium, interest rose in the automatic version as the ultimate representative of the twentieth century.

DAYTONA MILLENNIUM

There was hardly a more symbolic occasion for the introduction of a completely redesigned version of the Daytona than the beginning of a new millennium. The new model presented by Rolex in 2000 embodied—like the first Cosmograph thirty-seven years earlier—the chronographs of the future.

Externally, the new model deliberately borrowed from the 1988 Daytona. The innovations in the Oyster Perpetual Cosmograph Daytona of the new millennium were hidden within its case. The model had a fully integrated, specially developed new chronograph movement with a self-winding mechanism: the caliber 4130. This high-performance watch movement, full of innovative and patented technical solutions, set new standards for self-winding automatic chronographs not only in terms of robustness, reliability, efficiency, and precision, but also ease of maintenance.

The special performance of the caliber 4130 results primarily from the use of a vertical clutch to control the chronograph function instead of the classic horizontal clutch. The coupling of two superimposed discs enables an extremely precise, jerk-free start of the time measurement, and the force-free clutch makes it possible to operate the chronograph for a long time without affecting its precision.

In the caliber 4130, the engineers at Rolex have succeeded in reducing the number of components required for the chronograph mechanism by around 60 percent. In particular, they have significantly simplified the system of hour and minute totalizers—which traditionally consisted of two separate mechanisms on either side of the movement—by combining them into a single module with a decentralized clutch, cleverly housed on one side of the movement.

In 1988, the Cosmograph with self-winding automatic movement replaced the classic manual-winding Daytona. Initially, modified Zenith movements were used, and from 2000 onward (*center*) the company's own caliber 4130. Below is a version of the sports Daytona in a platinum case from 2013.

Oyster Perpetual Cosmograph Daytona

Reference: 116500LN

Movement: Self-winding, Rolex caliber 4130; 30.5 mm diameter, 6.5 mm height; 44 jewels; 28,800 vph; Parachrom hairspring, Glucydur balance wheel with Microstella regulating nuts; power reserve 72 hours; certified chronometer (COSC)

Functions: Hours, minutes, small seconds; chronograph

Case: Stainless steel; 40 mm diameter, 12.8 mm thickness; ceramic bezel (Cerachrom) with tachymetric scale; sapphire crystal; screw-down crown and pushers; water resistant to 10 bar

Bracelet: Oyster in stainless steel, folding clasp, with safety catch and extension link

Price: $13,122

The space thus gained was used to accommodate a larger mainspring and, in this way, to increase the movement's power reserve from around fifty to seventy-two hours.

A continuous, traversing bridge holds the balance wheel and the escapement securely in place, but in fact, one of the most spectacular of the newly introduced innovations was the Parachrom Breguet hairspring. This hairspring, developed and patented by Rolex, manufactured in-house, and made of a niobium-zirconium alloy, is absolutely insensitive to magnetic fields and has proven itself to be extremely stable regarding temperature fluctuations; beyond this, it works with up to ten times more precision than a conventional hairspring.

THE YOUNGEST GENERATION

The characteristic tachymeter bezel on the new Daytona presented in 2016, with (unchanged) Reference number 116500, is made from a solid block of the high-tech ceramic Cerachrom. Due to its hardness, now not only is the bezel extremely scratch resistant, but also the color is UV resistant and it is corrosion resistant. In addition to its extreme longevity, the numerals on the tachymetric scale are extraordinarily easy to read. First, the graduated scale is formed on the ceramic material and then a fine platinum layer is vapor-deposited on it, using the PVD (physical vapor deposition) process. This exclusive component was first used in 2011 on a Daytona model in rose gold

("Everose" in the Rolex nomenclature), then was used in 2013 to adorn the platinum model with a swimming-pool-blue dial, to mark the Daytona's fiftieth birthday.

The rest of the Daytona's Oyster case, waterproof to a depth of 100 meters, remained unchanged.

The new Daytona has "superlative chronometer" certification—a precision standard that was redefined by Rolex in 2015. This exclusive rating certifies that the watch has passed a number of tests that Rolex has carried out in its own laboratories and according to its own criteria. The certification tests are performed on the fully assembled watch and thus guarantee superlative performance for daily use in terms of reliability, water resistance, automatic self-winding, and power reserve. The rate tolerance of a superlative Rolex chronometer after it has been encased is plus or minus two seconds per day; thus, the criteria for precision when the watch is worn are more than twice as strict as those that must be met for official certification as a chronometer. The green seal affixed to every Rolex watch is associated with an international five-year guarantee.

Peter Braun

For decades, the winner of the Daytona twenty-four-hour race has received an engraved gold or stainless-steel/gold Rolex Daytona along with their trophy.

ROLEX
OYSTER PERPETUAL
SUPERLATIVE CHRONOMETER
OFFICIALLY CERTIFIED
SWISS
MADE
YACHT-MASTER II

Countdown

for the Advanced

The Rolex Oyster Perpetual Yacht-Master II, with its countdown timer that can be programmed in one-minute increments, was promoted as being the men's toy of the year in 2007. The "normal" Yacht-Master has been propagating a new look for water sports watches with rubber elements appropriate for sportswear since 2015.

The Yacht-Master II is equipped with the caliber 4161, newly developed in 2007; with its 360 components, it gives watch connoisseurs an inkling of how complex the mechanism is. Like the Daytona Chronograph caliber 4130 movement, this one has a column wheel and a vertical clutch. The heart of the watch movement beats at 28,800 vph and is equipped with a blue Parachrom Breguet hairspring from Rolex, reportedly more than ten times more resistant to vibrations and less sensitive to magnetic fields. It goes without saying that the movement is checked for precision by the independent Swiss testing body, the COSC (Contrôle officiel suisse des chronomètres).

The down counter can be preprogrammed in increments of one minute (up to a maximum of ten minutes) and is operated using the two pushers and the "Ring Command" bezel, which can be rotated unidirectionally by 90 degrees. Essentially, you can make the settings single-handed while wearing the watch, but to do this you first have to understand the system. Turning the Ring Command bezel blocks the start-stop pusher (at the "2") and enables an additional crown function at the same time. When you press the reset pusher (at the "4") and it clicks into place, the winding crown in position 1 (normally "winding") acts directly on the countdown minute hand. This lets you set a countdown time in minute increments. After turning the bezel back to its original position, the reset pusher pops out again and the crown can be screwed down to become waterproof. When the start-stop pusher (now unlocked again) is pressed, this initiates the countdown measurement. Of course, you can interrupt—and continue—the countdown at any time by pressing the start-stop pusher again. When the countdown is stopped, pressing the reset pusher sets the countdown to the last preset time. The highlight: While the ongoing countdown is in progress, the reset pusher works like a flyback; that is, the second hand jumps to zero and the minute counter to the previous or next minute, depending each time on the position of the seconds hand. In this way you can synchronize an already started countdown according to an official optical or acoustic signal.

The Yacht-Master II watchcase has a diameter of 44 mm and is made from a solid block of metal—stainless steel or yellow or white gold. The bezel of the yellow gold model is fitted with a blue ceramic dial, which bears the—quite striking—inscription "Yacht-Master II." The bezel for the white gold model, on the other hand, is made of platinum and displays the numerals in relief.

YACHT-MASTER WITH OYSTERFLEX BRACELET

The bracelet on a Rolex is an important component of the watch, and it is generally given a name. The link bracelets are called the Oyster, President, or Jubilee, and in 2015 the Oysterflex was added, a patent-pending rubber bracelet with a core consisting of fine metal blades made of nickel and titanium. These give the bracelet stability and prevent it from getting cut, while not impairing its flexibility in any way.

Oyster Perpetual Yacht-Master II

Reference: 116681

Movement: Self-winding, Rolex caliber 4161 (basic caliber 4130); 31.2 mm diameter, 8.05 mm height; 42 jewels; 28,800 vph; Parachrom hairspring, Glucydur balance wheel with Microstella regulating nuts; power reserve 72 hours; certified chronometer (COSC)

Functions: Hours, minutes, small seconds; programmable regatta countdown with memory

Case: Yellow gold; 44 mm diameter, 11.8 mm thickness; bezel with ceramic insert, bidirectionally rotatable; sapphire crystal; screw-down crown; waterproof to 10 bar

Bracelet: Oyster in yellow gold, folding clasp, with safety catch and extension link

Price: $43,599

Other versions: In rose gold ($25,179) and stainless steel ($18,530)

The fine-meshed metal core is encased in an elastomer that is flattering to your skin when you put it on. The well-shaped "cushions" also contribute to this. The bracelet is not adjustable, nor can it be shortened, which is why Rolex offers a total of six different bracelet lengths for the two watchcase sizes (37 and 40 mm), all of which can be combined with a finely adjustable folding clasp. Initially, this bracelet was exclusively available on the Yacht-Master, and that was in fact on the model in rose gold, which Rolex calls "Everose." The unidirectional rotating diving ring is likewise made of Everose, and the insert from the ceramic material Cerachrom.

The 40 mm Yacht-Master is powered by the Rolex in-house caliber 3135, which is equipped with the in-house blue-toned Parachrom Breguet hairspring and, like almost all Rolex watches, is certified as a chronometer by the COSC.

Martin Häussermann; pictures by Jörg Hajt

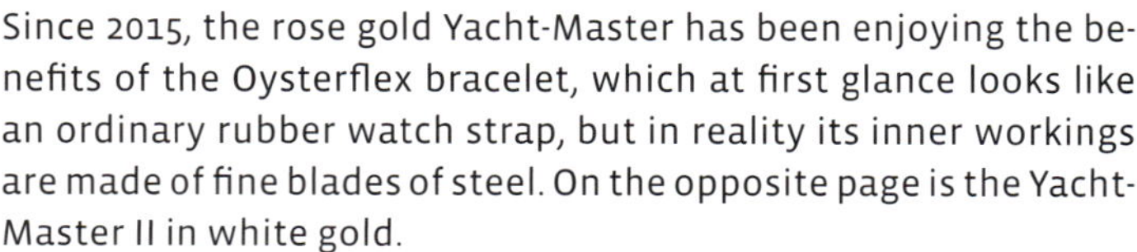

Since 2015, the rose gold Yacht-Master has been enjoying the benefits of the Oysterflex bracelet, which at first glance looks like an ordinary rubber watch strap, but in reality its inner workings are made of fine blades of steel. On the opposite page is the Yacht-Master II in white gold.

ROLEX

For **Fliers** and **High Fliers**

In 2012, Rolex also equipped its Sky-Dweller model with the rotating bezel called the "Ring Command," which can be used to conveniently control additional functions. To this, it is used to assign a range of tasks to the crown.

GMT watches or world time watches can be adjusted—preferably easily and quickly—to your new local time after a flight and yet always let you keep the time at home in mind. This is practical, and obviously also something popular with customers; otherwise this function would not enjoy its steadily growing popularity.

Rolex already had functional time zone watches in its portfolio for decades with the GMT-Master II and the Explorer II, but in the meantime there is now a more comfortable and modern way.

FOR DWELLERS OF THE SKY

The Sky-Dweller model was presented in 2012—referencing the Sea-Dweller diving watch—and the name probably refers equally both to the watch and to its wearer. Rolex developed the new caliber 9001 movement for the Sky-Dweller; this caliber also offers an annual calendar besides the second time zone. This watch can distinguish between months with thirty or thirty-one days and always switches to the first day of the next month at the correct time. The only time it is necessary to correct the calendar manually is at the end of February, and this has to be done for several days, because in the Rolex calendar February has thirty-one days.

An essential control element on this watch is the bidirectionally rotatable fluted bezel, called the "Ring Command" bezel. The wearer uses it to select which display he wants to change and then sets this using the crown.

In detail, it works like this: First we unscrew the crown and pull it out. Nothing happens yet. Then we turn the bezel and at the same time feel for the four stop positions, which takes a bit of practice at first.

Let's turn the bezel counterclockwise to the fourth and last position: Here we start the process for setting our home or reference time, represented by the twenty-four-hour disc and the minute hand. The stop seconds function, which we need to set the watch precisely to the right second, also works in this position of the bezel.

The twenty-four-hour disc on the Sky-Dweller dial shows the time at home; we can read this at the red arrow marker.

ROLEX
OFFICIALLY CERTIFIED
SWISS

Then move the bezel clockwise by one notch. The seconds hand starts moving again. In this position we can set the local time by using the crown to set the position of the hour hand. Air travelers will be happy to use this function after a change of time zone, especially since it isn't necessary to stop your watch to do this. It can go on running just as precisely as a Rolex runs.

Last but not least, we set the day and the month. To do this, we carefully turn the bezel once again in a clockwise direction, one notch farther. Using the crown, we now move the day forward until the month indicator jumps to the correct position. The uninitiated will discover this only at a second or third glance, because this does not involve a display with the names or abbreviations for the months. Rather, Rolex uses the Roman numerals on the time display to indicate the month. This is done when the small window above each numeral—I for January, II for February, etc.—gets filled in in black, according to the current month.

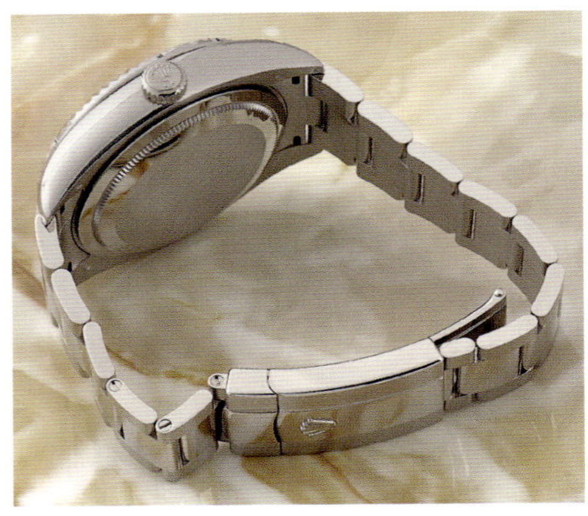

The month display for the annual calendar uses the twelve numerals for the hours or the twelve small display windows next to them. *Below*: The Rolex in-house watch movement is showcased in its straightforward design and impeccable workmanship.

Now we set the date and turn the bezel back to its starting position—done. Then we can make any necessary corrections at the respective setting positions. When it is screwed down, the crown on the winding crown should always be perpendicular to the plane of the dial—this, by the way, is also patented.

A COMPLICATED INTERNAL LIFE

As with all the newly launched Oyster models in recent years, the blue Parachrom hairspring with the Breguet overcoil, which Rolex developed itself and likewise manufactures itself, is also used in the new caliber 9001. According to information from Rolex, the alloy used to make it is completely nonmagnetic, and it is significantly more shock resistant than other commercially available hairsprings. These watches are designed for everyday wear because of such features as the in-house "Paraflex" shock absorber and water resistance to 10 atm (water pressure at depths of 100 meters). Like all mechanical watches made by Rolex, the Sky-Dweller is chronometer-tested.

For the Sky-Dweller, Rolex limits itself to a 42 mm case that is around 14 mm thick. This way it doesn't look oversized on comparatively slim wrists, but it doesn't shine with exaggerated restraint either. And it is very comfortable to wear, thanks to the perfectly crafted solid Oyster bracelet. Its folding clasp is milled from a solid piece to inspire confidence, and will open only as the wearer wishes.

Martin Häussermann; pictures by Jörg Hajt

Oyster Perpetual Sky-Dweller

Movement: Self-winding, Rolex caliber 9001; 33 mm diameter, 8 mm height; 40 jewels; 28,800 vph; power reserve 72 hours; certified chronometer (COSC)

Functions: Hours, minutes, center seconds; additional 24-hour display (second time zone); annual calendar with date and month

Case: White gold; 42 mm diameter, 14.1 mm thickness; bidirectionally rotatable bezel to control the functions; sapphire crystal; screw-down crown; waterproof to 10 bar
Bracelet: Oyster white gold with folding clasp.

Price: $46,668

Other versions: in rose gold; in stainless steel ($14,900); with Oysterflex elastomer bracelet (in gold, $40,000)

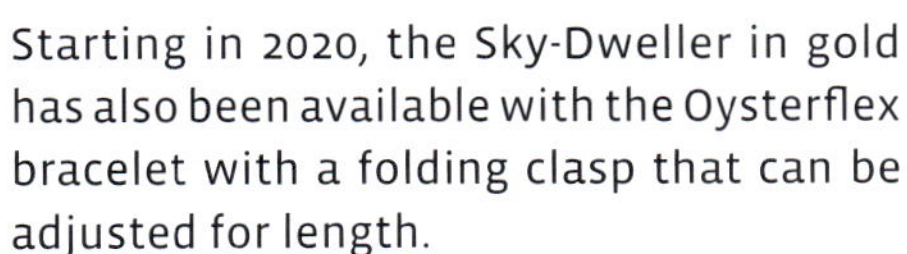
Starting in 2020, the Sky-Dweller in gold has also been available with the Oysterflex bracelet with a folding clasp that can be adjusted for length.

ROLEX
GENEVE
SWISS MADE

The Moon at the Sign of the Crown

With the Cellini Reference 50535, the moon phase is celebrating its glamorous comeback at Rolex. This visually attractive complication was made only in small numbers during the 1950s and had disappeared from the brand's catalogs since 1960.

It felt like an eternity after the last (and only) Rolex Oyster model with what was called an "astronomical display" at the time disappeared from store windows due to lack of demand. Reference 6062, with full calendar with moon phase, achieves astronomical prices in collectors' markets due to its sheer rarity. From a collector's point of view, it is therefore almost unseemly to wish the new Rolex Moonphase watch any significant success.

EXCEPTIONAL COMPLICATIONS

In the more than one hundred years of its existence, the Rolex brand has primarily been making simple, technically robust three-hand watches with self-winding movements, with and without date or day-of-the-week display; sports watches; and elegantly packaged watches in waterproof cases and fitted with metal link bracelets with folding clasps. The proportion of chronographs in the collection always remained well below the 10 percent mark, while the company even continued to purchase their blank movements up until 2000. Watches with complications—and this includes timepieces with advanced calendar functions—had no role to play in the Rolex universe during the last century.

That has changed since 2000. After the indicated in-house chronograph caliber (4130) for the Cosmograph Daytona, engineers designed a regatta counter with presettable countdown (caliber 4161, Yacht-Master II model) and an annual calendar with second time zone (caliber 9001, Sky-Dweller model).

Interesting small complications have also been introduced in the Cellini line; for example, a pointer date display (caliber 3165, Cellini Date model) or a second time zone display (caliber 3180, Cellini Dual Time model). So in a way, the moon phase display has been in the pipeline for quite some time.

PREMIERE OF THE MOONPHASE

The Cellini Moonphase takes its place at the head of the elegant-model family, which has been thriving in the shadow of the overwhelming Oyster collection since it was revitalized in 2014. The caliber 3195 is a completely in-house design with a drive integrated into the wheel train bridge for the moon phase display. This means that it is not only moved once a day by the date mechanism but also is set permanently in the power flow of the watch movement and thus has achieved the incredibly low rate deviation of just one day in 122 years. Over the months, simply switching the twelve- or twenty-four-hour rhythm will lead to the accumulation of a considerable deviation from the actual cycle of the moon. This comprises twenty-nine days, twelve hours, forty-four minutes, and three seconds and thus is not divisible by 12 or 24.

It became a bestseller only fifty years after its end: the Rolex Oyster Perpetual Reference 6062 with moon phase display.

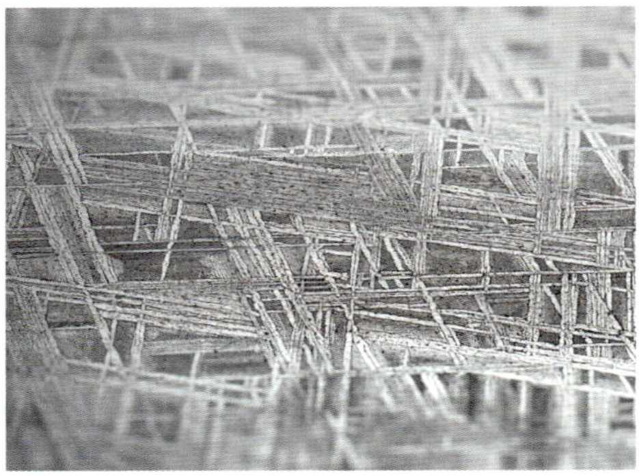

It takes some getting used to: This moon doesn't disappear behind the edge of a cloud but instead revolves in the open at double the speed. It is read at the index marker. The surface of the moon disc is made of polished, etched, and rhodinated meteorite rock.

The independent drive for the moon disc on the Rolex Cellini Reference 50535 dial makes it possible to create a distinct interpretation for the presentation: A full-moon disc made of rhodium-plated meteorite gleams next to decorative stars against an enameled background of midnight blue. The disc rotates completely once per moon phase, with the symbols for the full moon and the new moon visible the entire time; they do not just appear temporarily in a window opening, as is usual in other moon phase displays.

The current phase of the moon is indicated by a small, golden, triangular index on the upper rim of the disc display. This means you can get only an approximate reading of the phase of the moon, although this has no relation to the accuracy of the mechanism itself—the conventional moon phase display on the Rolex Oyster Reference 6062, which was made between 1950 and 1960, managed this much better.

In addition to the moon phase display, the Cellini Moonphase also features a long, blue-colored center date hand, with its crescent shaped tip sweeping over a thirty-one-day scale on the outermost edge of the dial. Incidentally, the date is not preset for 122 years, but rather it is necessary to correct it manually at the end of the month for February, April, June, September, and November.

CHRONOMETER OF THE SUPERLATIVE

The Cellini Moonphase is available only in rose gold ("Everose") with a brown alligator leather strap with a concealed Crownclasp folding clasp. The latter, by the way, is another first for a Cellini model, because the Date, Time, and Dual Time model straps all have pin buckles.

Like all Rolex wristwatches, the Cellini Moonphase is also certified as a "superlative chronometer" (the new precision standard from Rolex). According to the 2015 definition, the tests performed by the COSC on dismantled movements are no longer sufficient, so Rolex puts each finished watch back on the testing bench and tests it according to its own internal criteria. These are stricter than the watch industry standards: The rate tolerance of a superlative Rolex chronometer after it is encased is plus or minus two seconds per day. That is only half as much as is tolerated for an official certification as a chronometer according to COSC. In addition, the watches are tested for their winding performance, robustness, and water resistance.

On the elegant Rolex Cellini Moonphase, it is just as difficult to get a glimpse of its ability to roll with the punches as it is to see its precision.

Cellini Moonphase

Reference: 50535

Movement: Self-winding, Rolex caliber 3195; 28.5 mm diameter; 31 jewels; 28,800 vph; Parachrom hairspring, Paraflex shock absorber; power reserve 48 hours; certified chronometer (COSC and Rolex)

Functions: Hours, minutes, center seconds; date, moon phase

Case: Rose gold, 39 mm diameter, sapphire crystal, screw-down crown, waterproof to 5 bar

Bracelet: Reptile leather, rose gold folding clasp

Remarks: Blue-enameled sky disc with moon made of rhodium-plated meteorite

Price: $26,696

ELEGANT SIBLINGS

As already mentioned, the Cellini Moonphase is just the icing on an already very appetizing cake. The Cellini line is based on a slim 39 mm gold case with a cambered and finely fluted double bezel, which, if desired, can also be set with a row of diamonds. The screw-down case back is classically domed and hides the technology of the tried-and-true self-winding caliber family 3100 movement from view. Caliber version 3132 is encased in the Cellini Time model with center seconds, which is already equipped with the nonmagnetic blue Parachrom hairspring. The Date version of this model has a pointer date display at the "3," and the corresponding self-winding movement is caliber 3165. The Dual Time version is equipped with a second time zone display that can be set by the caliber 3180 winding crown, completely independently of the central time display, and has a day/night indicator. When traveling, you can keep your home time on the small auxiliary dial at the "6" and see at a glance whether it is currently day or night at home. This dial is also fitted with hour and minute hands, making it easier to read.

While the modest Time model gleams with its smooth white or silvered dial, the two versions with additional displays dazzle with their radial guilloche dials. The appliqué hour markers as well as the Rolex crown are always made of white gold. ♛

Peter Braun

Left row, from the top: Cellini Time in rose gold ("Everose") and white gold; *below,* the Dual Time in white gold. *Top right,* the Time design in white gold with a diamond bezel; *below,* the Date in white gold and the Dual Time in "Everose."

V.
The Luxury Watch Manufacturer

The Coronation

During a visit to the four manufacturing locations in Geneva and Biel, we were able to gain deep insights into the structure and functioning of the largest luxury watch manufacturer in Switzerland.

ROLEX
OYSTER PERPETUAL
DAY-DATE

Four great classics: the Rolex Oyster Perpetual Yacht-Master II, the Cosmograph Daytona, the GMT-Master II, and the Submariner Date

Rolex is a phenomenon. No other European luxury watch manufacturer sells so many—exclusively mechanical—watches every year, and no other watch brand has experienced such good development during the last century and established itself so successfully during the past decades. In the past 112 years, the Rolex brand created by Hans Wilsdorf has become an icon, and today it is probably one of the most profitable watch manufacturers in the world. This undoubtedly has been a matter of successful communications right from the start—for example, with a young swimmer who swam across the English Channel wearing a Rolex in 1927 and found herself on the front page of the *Daily Mail* the next day. But also a matter of communications with statesmen, athletes, and movie stars who show off their Rolexes in public as a matter of course.

At least as important for such success is the company's high level of independence. This applies to its policy for models and its many-times-copied designs, but above all to its technical know-how. Rolex is the watch brand with probably the highest-level vertical range of manufacturing in Europe. As a foundation freed of pressure from shareholders and the greed of individual owners, Rolex was able to—indeed, it had to—put the profits generated back into the company, over and over again. In addition to modern production machinery and spacious buildings, in recent decades the company has preferred to invest in the purchase and integration of what formerly were its external suppliers.

Today, this giant simply designs and manufactures everything in-house, apart from the leather straps, watch hands, and sapphire crystals. For some years now Rolex has also belonged to the illustrious circle of watch manufacturers that are able to produce their own balance wheels. And it is not doing this in restricted quantities: in the recent past, the number of watches Rolex has been producing annually is likely to have ranged between 700,000 and 800,000. There are in fact no official figures from Rolex because, as a foundation, the company is obliged to publish only limited information. Because Rolex is now submitting all of its movements to the Official Swiss Chronometer Testing Institute (COSC) for testing—and in fact to their branch in Biel—it is possible to arrive at halfway realistic figures at least.

This means that a Rolex watch, seen quite objectively, is a mass-produced item, yet because the supply on the markets is always kept at scarcity levels, those models that are in particularly high demand become genuine luxury goods—made of steel! Rolex is also a popular theme among collectors, not least because of its high level of value retention. All this is hardly a matter of coincidence, as our editorial team members Peter Braun and Martin Häussermann found out earlier this year during a visit to all four Rolex manufacturing locations in Geneva and Biel.

Rolex in Les Acacias:
COMPANY HEADQUARTERS AND FINAL ASSEMBLY

Let's start at the headquarters. Since 1965, Rolex's fortunes have been determined in the Les Acacias district of Geneva. The building complex, which was completely modernized between October 2002 and October 2006, looks very impressive from the outside, as well as from the inside. It appears to us, so to speak, as a cipher for the entire brand. With its six floors, it not only rises high up but also goes deep down to four basement floors.

As in the other two manufacturing facilities in Geneva, these ultramodern buildings are dominated by facades of glass and steel. The slender high-rise administration building, with a green-glass facade—at Rolex, the manufacturing facilities feature smoke-black glass surfaces—is not only where all the central company decisions made, but it is also the location where sponsoring activities are directed and distribution of the watches is organized. This location is also the home of the design and research-and-development departments; the same is true for customer service, as well as assembly and final inspection.

Clinical cleanliness is the top priority when assembling watches. It is possible to enter the premises only through air locks and when wearing special shoes—as visitors we are given shoe covers—and lab coats. After all, work is done here under clean room conditions. Permanent overpressure in the production hall ensures that it is impossible for any harmful dust to penetrate the watches.

The Rolex headquarters in Geneva Les Acacias. The slender high-rise building has been the landmark at the site since 1965; it was completely renovated between 2002 and 2006.

At Rolex, administration buildings are glazed in green and production facilities are glazed in black.

Common features of all Rolex manufacturing facilities include their generous amount of space and their meticulous state of cleanliness.

What always remains impressive is the large proportion of work done by hand, both in assembly and inspection, despite the high level of industrialization.

Assembly of the watches is organized as group production. The teams of twelve to fourteen people each work completely independently. At the beginning of the week, the teams are given their weekly work target and then decide who is to perform which assembly steps. Working in this way, the team members avoid monotony despite the extreme division of labor in practice and gain experience working with all caliber types. This contributes to a high quality of manufacturing. Up to eight different assembly processes are performed at each workstation.

Although there is a total of 150 people sitting here at the long work tables on two floors, a concentrated silence prevails. There is not much to be heard, with the possible exception of the hum of the small electric screwdrivers, which are of course always set to the tightening torque required in each case. Incidentally, it requires exactly three Newton centimeters (cm) to assemble the automatic bridge.

The watch movements are delivered in dustproof plastic containers; from Biel, in fact, where Rolex manufactures all its movements without exception and then delivers them to the COSC, located in the same town—to be tested without the automatic self-winding mechanism engaged. It goes without saying that all the Rolex transport containers are labeled with barcodes. Not only does this make it possible to trace every movement and watch as it progresses through the manufacturing process, but the watch manufacturer's entire logistics system is based on these barcodes.

In the basement, the fully assembled watches (without bracelets) go through an eight-stage quality review. In 2015, Rolex introduced an in-house testing process that prescribes standards than are stricter than the usual industry standard. This process involves the performance qualities for precision, power reserve, the performance of the automatic self-winding mechanism, and water resistance. The rate precision is tested twice over. As already described, the finished movements will be tested by the COSC and given a certificate from this independent testing institute. However, because it is possible that something might change during assembly, Rolex also tests the assembled watches again, and in fact more strictly than the COSC has done. While the latter accepts deviations of −4 to +6 seconds a day, Rolex watches must comply with a tolerance of −2 to +2 seconds a day. This is tested on the one hand in a cycle in seven static positions lasting around fourteen hours, and otherwise in a dynamic process performed by an applicably programmed robot.

They are also strict when it comes to water resistance. In fact, every single watch is tested, initially with air pressure and later in the water. During the test, for watches that are nominally supposed to be waterproof to 100 meters (10 bar), a margin of 10 percent is added for the sake of security. All diving watches are tested with an added margin of 25 percent. Therefore, this standard applies for the Submariner (300 m / 37.5 bar), the Sea-Dweller (1220 m / 155 bar), and the Deepsea (3900 m / 495 bar).

Rolex in Brief

1905 Hans Wilsdorf from Kulmbach, in Bavaria, Germany, founds a company in London specializing in the watch trade.

1908 Wilsdorf invents the name "Rolex" as the signature for his creations.

1920 Wilsdorf founds Montres Rolex SA in Geneva; the watch movements for this company are supplied exclusively by the Biel-based manufacturer Jean Aegler.

1945 Wilsdorf, who was childless, donates all his shares in the company to his foundation.

1960 Hans Wilsdorf dies; his longtime associate André Heiniger directs the company.

1992 André Heiniger appoints his son Patrick as CEO.

2002 The production facility in Chêne-Bourg is opened (for setting gemstones and manufacturing dials and ceramics).

2004 Rolex Geneva purchases the Rolex Biel company from the Borer family (descendants of Jean Aegler).

2005 The manufacturing facility opens in the Plan-les-Ouates industrial area in Geneva.

2006 Rolex produces 613,000 tested chronometers.

2008 Bruno Meier becomes CEO. In 2011, Riccardo Marini succeeds Bruno Meier as the new CEO; the latter leaves the company.

2014 Jean-Frédéric Dufour, formerly CEO of Zenith, becomes the new CEO of Rolex.

2015 Rolex produces around 795,000 tested chronometers, according to COSC figures.

2019 Rolex is employing over 7,000 people in Switzerland.

In 2012, Rolex moved its watch movement production into a brand-new building ensemble on the northern edge of Biel, Switzerland.

Despite the automation of component production, there is still a lot for the well-trained staff to do: all watch movements are assembled, adjusted, and tested by hand.

Manufacture des Montres Rolex:
THE CALIBER SMITHS

Exactly fifteen years ago, an important decision was made in Geneva. Rolex purchased Manufacture des Montres Rolex SA, Biel, the watch movement factory that had been legally and economically independent until then, and integrated it into Rolex SA, Geneva. In this way they were killing two birds with one stone: gaining control over an existentially important supplier and securing what is arguably the company's most important export market. Not only was Rolex Biel manufacturing all the watch movements for the company, but Rolex Biel also held the trademark rights for the United States at the time. The official statement from Geneva at the time read: "This amounts to an epoch-making step that will strengthen the brand over the long term." There's probably no disagreeing with that.

A lot has happened in Biel since then. The new building, constructed in 2012 on the northern edge of the city, has increased the production premises to a current area of 135,000 square meters, and there are more than 2,000 people working here. Production is strictly organized according to manufacturing processes.

Traditionally, the largest use of machinery is for manufacturing the incomplete watch movements—without mainspring, dial, and hands—which are called "ébauches" at Rolex. These comprise the bottom plates, bridges, and cocks made of brass; calendar discs are also added to them. All the holes for the pinion shaft are bored in a single operation—a proven process for ensuring consistent axle spacing, which is an important prerequisite for the wheel train running smoothly and thus ensuring the highest possible rate precision.

Fully automated machining centers then carry out a large number of machining steps to mill and grind the still-rectangular brass plates into the correct shape. The greatest number of the machines comes from Précitrame, a specialist company in which Rolex has a financial stake. All the parts of the ébauche are visually inspected and randomly remeasured to detect and replace worn tools in good time, for example.

Rolex is one of the few watch manufacturers to have developed the know-how for making hairsprings—the component that makes a significant contribution to the precision of a watch. Manufacturing its own hairsprings is a

strategic advantage for a watch manufacturer that should not be underestimated. The so-called "Parachrom" hairspring was developed in the research department in Les Acacias, and initially was also manufactured there. In the meantime, production of the hairsprings was sensibly relocated to the watch movement manufacturing site.

Rolex mixed the material used to make its hairspring itself; it contains niobium and zirconium and initially appears as an inconspicuous metal rod about 30 cm long. Not only is the alloy of the basic material patented, but also the special manufacturing method—Rolex has developed a machine that is unique in the world to perform this process. In a rough description, this involves a vacuum chamber in which several rods made of different materials are heated, twisted together, fused, and cooled again with the aid of high voltage (5,000 volts). As a result of this treatment, the alloy becomes completely paramagnetic, and the influence of magnetic fields consequently has no effect on the precision of the watch. The protective oxide layer already mentioned is also formed here. A total of twenty-five rolling and drawing processes turn the 30 cm long rod into around 3 kilometers of hairspring, which is exactly 50 microns (0.05 millimeters) thick. It is significantly thinner than a human hair, but in contrast to a hair, it is the same thickness from front to back, enabling it to perfectly regulate the precision for which Rolex watches are renowned.

In the vacuum furnace, the Parachrom alloy is melted from bars of various basic materials.

Hairsprings and balance wheels are classified and sorted according to their moment of rotation or moment of inertia.

The automatic-winding group, with rotor, reversing wheel, and gear train

The Paraflex shock absorber for the balance wheel

Complete balance wheel with blue Parachrom hairspring

Rolex in Plan-les-Ouates:

MANUFACTURING WATCHCASES AND LINK BRACELETS

Three giant black monoliths compose the production center for the watchcases and link bracelets.

In Plan-les-Ouates, everything revolves around metalworking. Bezels are typical of turned parts; the raw material comes in the form of a tube.

Thousands of cases, bracelet links, and clasps are circulating through production during the day and are safely stored in the fully automated high-bay warehouse at night.

In the same neighborhood as Patek Philippe, Vacheron Constantin, and Piaget, Rolex produces watchcases and metal bracelets in three buildings on eleven floors—five of them underground. This includes not only assembly but also development, design, and quality control. Rolex staff—dressed in helmets and silver fire-retardant jackets—also melt down the gold needed to make the cases and bracelets in the company's own furnace. This includes the so-called Everose, a patented rose gold alloy with a platinum component. In this mixture, the reddish hue that is created by the copper it contains will endure permanently, while commercial alloys (with silver admixture) will actually turn yellow over the course of many years.

The Geneva-based company does buy the stainless steel it uses; for example, from the Austrian manufacturer Böhler. Of course, this is the highest-quality steel, designated as number 904L, which is also used to manufacture surgical instruments. Many watch manufacturers claim that they use this top-quality steel, but at Rolex they emphasize the fact that they searched for a long time to find the right manufacturer that had the capacity to deliver the desired quality. The alloy number does not provide any comprehensive information about the properties of the material. This is probably one of the reasons why Rolex maintains a sixty-strong research-and-development department at its Plan-les-Ouates manufacturing plant—and please take note that this is just for case and bracelet production!

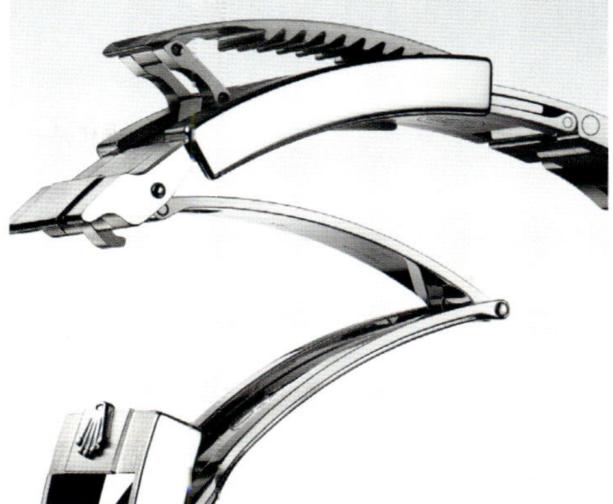

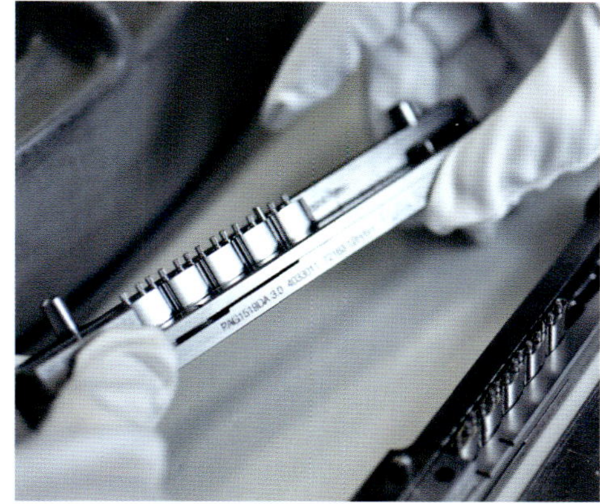

Polishing the Oyster case requires experienced fingers and a trained eye.

The Glidelock clasp is a complex precision component.

The bracelet links are sorted by hand and pressed or screwed together.

The crown of creation: the President link bracelet with concealed clasp

The view into the underground high-bay warehouse, as it is set up not only in Plan-les-Ouates but also in the three other manufacturing centers in Les Acacias, Chêne-Bourg, and Biel, is also impressive. Controlled by a sophisticated enterprise resource planning system (SAP), all workpieces are locked away here every evening at the end of the shift, including the raw materials, semifinished products, and finished cases and bracelets. The next morning, before work begins, fully automated robots sort the workpieces to their places in the individual departments. The warehouse robot, which whizzes through the narrow aisles of the high-bay warehouse (60,000 storage locations in Plan-les-Ouates) at amazing speed and executes up to 2,800 movements per hour, transfers the parts, packed in standardized carrier boxes, throughout the building via a "pneumatic tube" system. Above the suspended ceilings in the corridors, you can hear the electrically driven roller conveyor tracks whirring quietly, and in some departments, self-propelled transport robots, lights blinking—jokingly called "R2-D2" by the staff—make their way to the loading side of the machines.

The Rolex Training Center

The newly equipped training center for education and further training of staff is in the immediate vicinity of the company headquarters, in the center of the Les Acacias district. The facility is spread over two floors, comprising a total of 4,300 square meters. Year after year, nearly 3,000 management personnel are trained here in courses on product knowledge, professional expertise, and management. Rolex was the first employer in the watchmaking industry in Geneva to offer in-house training, with the creation of apprenticeships in 1984. Today, the second floor of the training center is used to teach horologists/watchmakers, polymechanics, and the skilled staff who, under the Swiss system, spend half of their training period on the job. The training center supervises about a hundred apprentices who are in different years of their apprenticeships. While training to become a skilled worker (a "watchmaking operator" in international English usage) takes two years, a production watchmaker has to spend three years and a "full watchmaker" requires four years at school.

Rolex in Chêne-Bourg:

WATCH DIALS, CERAMICS AND STAINLESS STEEL, AND SETTING GEMSTONES

The integration program began in 2002 with the purchase and subsequent conversion of a watch dial factory in the Chêne-Bourg district.

Electroplated dials ready to be fitted with the hour markers and logo

This dial is decorated with a fine radial sunburst pattern. The holes for the hands and the date window are made with the highest precision.

In 2002, Rolex acquired the watchcase manufacturer Genex, in the southern Geneva district of Chêne-Bourg, which lies directly on the border with France. Rolex subsequently developed its former supplier into a high-tech company, tailored to the specific needs of its own portfolio.

Brass plays an important role as a basic material for manufacturing the dials, while Rolex uses only gold for setting gemstones. The brass discs, rolled to size and ground flat, are prepared in several processing steps, regardless of how they will be used subsequently. This is done first on machine tools that punch out the windows for the date and day-of-the-week displays, then drill and trim the tiny holes for positioning the appliqué hour markers and the polished little crown. After that, the dial blanks are put through several galvanic baths that harden and stabilize the soft and coarse-pored surface of the raw brass. Finally, tiny mounting feet are soldered on the back of the dial blank, which are later inserted into corresponding holes in the movement bottom plate, or the rim is flanged: on the latest generation of movements, the dials are simply slipped into place; a groove running around the main bottom plate ensures an absolutely tight fit.

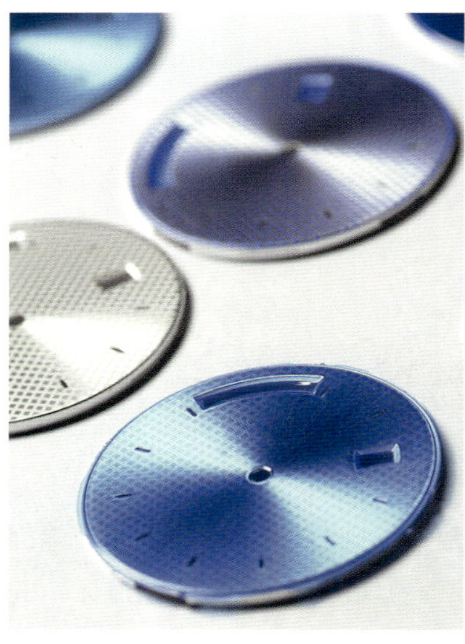

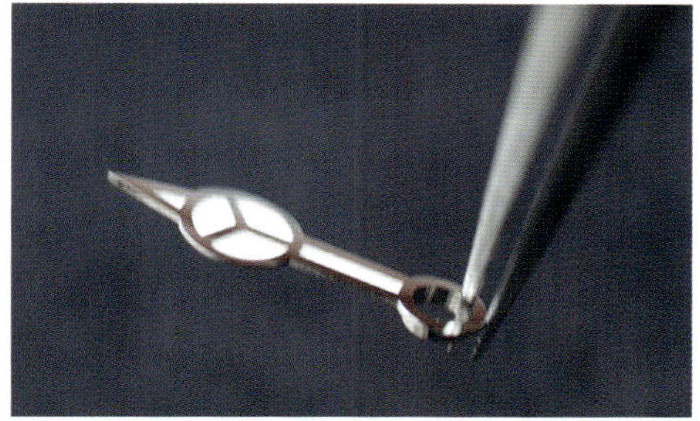

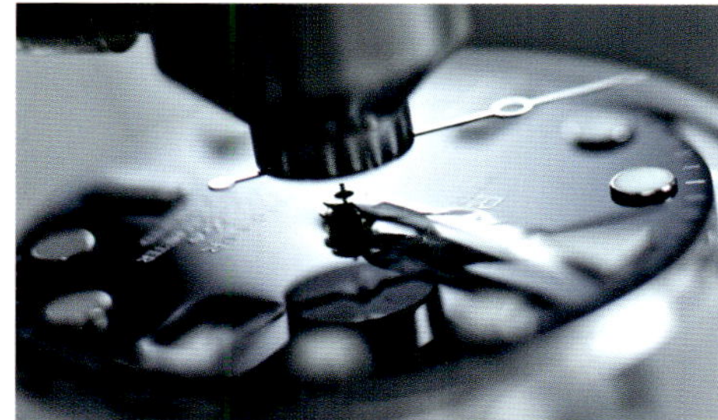

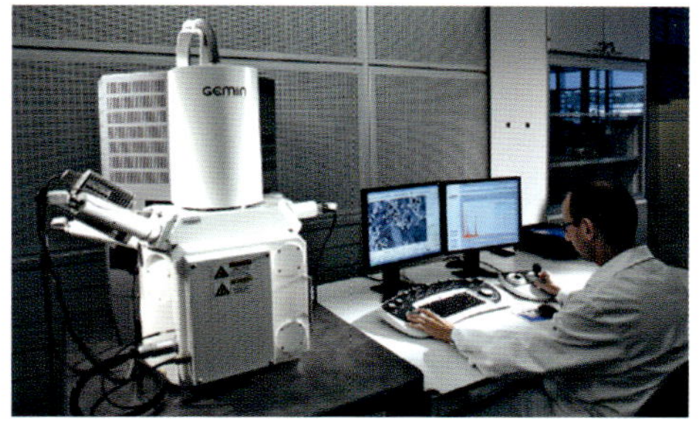

At this point, the methods used to manufacture the dials diverge, depending on how they will be used: some will be lacquered and others polished and galvanized. In the hands of the specialists—in fact, most of the staff in the Terminage or Finishing Department are women—they then all become the same again: the lettering is transferred from the template to the dial in several steps, using a printing tampon, and the appliqués— all of which, by the way, are manufactured in-house without exception—are threaded into the holes with their hair-fine feet first and riveted from behind. All of this work is performed by pure, concentrated hand workmanship, at best supported by small stamping presses or holding devices developed in-house.

Even more hand workmanship can be found in the gemstone setter's studio, where, as mentioned, only the gold dials for women's jeweled models will be set with brilliant-cut diamonds.

The manufacture of ceramic bezels made of the patented material Cerachrom, recently relocated from Plan-les-Ouates, is still quite new at the Chêne-Bourg manufacturing facilities. In a separate department, the powdery starting material, consisting of 50 percent zirconia and color pigments and 50 percent of a liquid bonding agent, is compressed and injected into molds under high pressure. The blanks, which are still brittle, are "debound" over two full days—that is, everything is evaporated out at temperatures around 1,290°F—and then the material is sintered under high pressure and at high temperature. In these processes the bezel rings shrink by 25 percent, because the newly arranged zirconium atoms and pigments are now moved closer together and more firmly bonded than before. After the sintering, it is no longer possible to use normal tools to make an impression on the Cerachrom material. Polishing the surfaces and deburring the edges become an elaborate feat of exerted force.

Martin Häussermann and Peter Braun

The dials, completed with appliqués, are mounted on the watch movement and fitted with the exactly positioned hands by staff working at partially automated workstations.

The testing laboratory in Chêne-Bourg is equipped with a scanning electron microscope.

Dials that are set completely with diamond pavé are a specialty of the house.

VI.
The Ten Outstanding Features of a Rolex

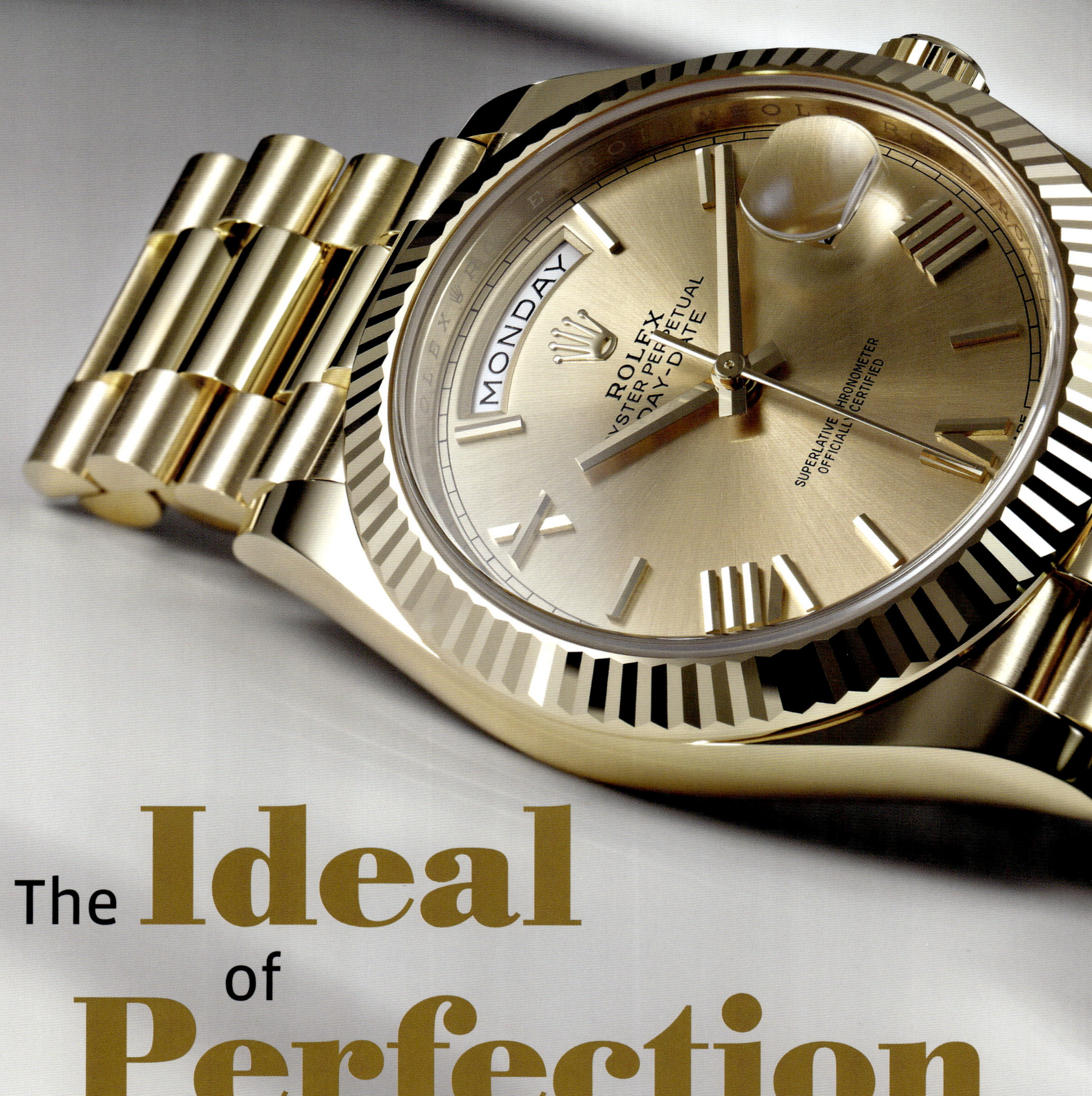

The Ideal of Perfection

Rolex achieved pioneering work in the development of the wristwatch. However, the watch brand has not rested on its laurels; instead, it has continued to refine all of its components, introduced new materials, and continually optimized its processes.

Some watch connoisseurs are of the opinion that it is no longer possible to make any more technical or functional advances in the wristwatch sector. After the invention of the waterproof case with screw-down crystal, case back, and crown, and the introduction of self-winding with an oscillating-weight rotor, the wristwatch had basically been completely developed. The Rolex Oyster Perpetual combined all of these achievements and thus was miles ahead of the competition in 1931. Internationally registered patents secured this lead until the 1950s, after which rotor winding, screw-down crowns, and screw-down case backs became standard throughout the watch industry. Evil tongues sometimes claim that after the introduction of the Datejust in 1945, Rolex abandoned making any kind of further refinement of its watches and is still producing the same two watch models up to today—a robust model in stainless steel with a rotating bezel or an elegant design in gold with a diamond-cut bezel. Of course that's not entirely true, but there is an obvious reason why Rolex has stubbornly resisted the trend toward complication watches: the stately army of Rolex engineers has put its innovative power at the service of perfection over the past decades, repeatedly formulating and implementing new standards.

Only genuine with the seal: the green tag with the inscription saying "Superlative Certified" is simply an integral part of every Rolex watch.

1 Superlative Chronometer Officially Certified: PRECISION ABOVE ALL

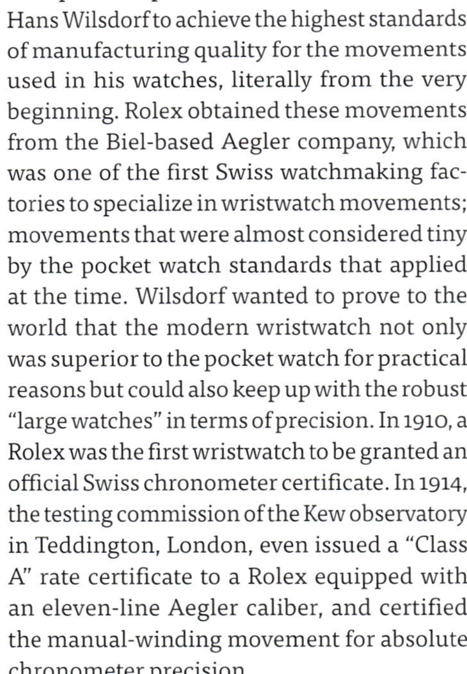

The quest for precision drove Rolex founder Hans Wilsdorf to achieve the highest standards of manufacturing quality for the movements used in his watches, literally from the very beginning. Rolex obtained these movements from the Biel-based Aegler company, which was one of the first Swiss watchmaking factories to specialize in wristwatch movements; movements that were almost considered tiny by the pocket watch standards that applied at the time. Wilsdorf wanted to prove to the world that the modern wristwatch not only was superior to the pocket watch for practical reasons but could also keep up with the robust "large watches" in terms of precision. In 1910, a Rolex was the first wristwatch to be granted an official Swiss chronometer certificate. In 1914, the testing commission of the Kew observatory in Teddington, London, even issued a "Class A" rate certificate to a Rolex equipped with an eleven-line Aegler caliber, and certified the manual-winding movement for absolute chronometer precision.

Since the end of the 1950s, most of the models in the Oyster collection have been designated as "superlative chronometers," because the movements of the caliber 1500 family exceeded the requirements of the official Swiss testing institute. However, it is only since 2015 that there has been a testing procedure, devised and performed in-house by Rolex, which has proven its outstanding rate values. The tolerances are set within an even-narrower range in comparison to the requirements of the COSC (Contrôle officiel suisse des chronomètres, or Official Swiss Chronometer Testing Institute): the average daily rate must be between +2 and −2 seconds (COSC: −4/+6 seconds).

As early as 1927, Hans Wilsdorf said: "We work according to criteria that can be measured only with instruments that we manufacture ourselves." Measurements are therefore carried out in a separate department in the basement of the Rolex production center in the Plan-les-Ouates district of Geneva, after the watch has previously passed the COSC's official test and in fact with the movement already engaged (but without the bracelet). In this fully automated "superlative" test procedure, the winding performance of the self-winding movement, the power reserve, and water resistance (with a safety margin of +25 percent) are also inspected.

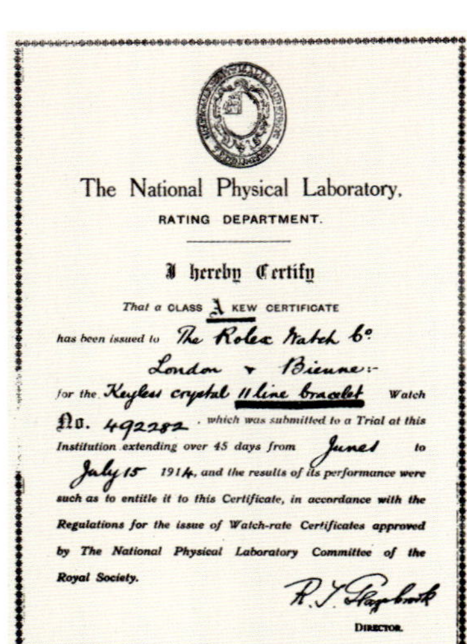

The National Physical Laboratory.
RATING DEPARTMENT.

I hereby Certify

That a CLASS A KEW CERTIFICATE has been issued to The Rolex Watch Co London & Bienne for the Keyless crystal 11 line bracelet Watch No. 492282 which was submitted to a Trial at this Institution extending over 45 days from June 1 to July 15 1914, and the results of its performance were such as to entitle it to this Certificate, in accordance with the Regulations for the issue of Watch-rate Certificates approved by The National Physical Laboratory Committee of the Royal Society.

R. T. Glazebrook
DIRECTOR.

Since the late 1950s, most Rolex watches have already been running more accurately than the COSC prescribes.

2 The Screw-Down Winding Crown: EFFECTIVE PROTECTION

The ever-practical Hans Wilsdorf saw the exposed place where a wristwatch is worn as a potential problem, because the potential for penetration by dust and moisture could wreak havoc on the delicate movement mechanism. He was not the first to recognize the crown tube as a "gateway" into the case interior, but his approach of screwing the crown down onto a threaded connection with an inserted seal was the most sensible and practical way to solve this problem.

The principle of the crown as the "cap nut" on the case has not changed much to this day. After the patent expired, Rolex improved the system and, with its "Twinlock" crown, once again secured itself with a leading edge over the competition following rapidly on its heels.

This system is used in all the wristwatches in the Oyster collection, which are guaranteed to be waterproof to a depth of 100 meters. The Twinlock system can be recognized—depending on the material used for the crown—by the presence of one point or two points, or by a line under the Rolex crown. The "Triplock" crown, introduced in 1970, has two sealed areas inside the crown tube and a third sealed area in the winding crown itself. This system was developed to increase the water resistance of the Submariner, Sea-Dweller, and Deepsea diving watches to 300, 1,220, and 3,900 meters, respectively. Today, it also is part of the equipment of various other Professional models. The Triplock crown is marked with three points.

Two points under the crown stand for the Twinlock system.

The Oyster Watchcase: CLAMPED SHUT LIKE AN OYSTER

Being waterproof is regarded as tried-and-tested proof of a hermetically sealed watchcase. At the same time, Hans Wilsdorf was originally concerned first and foremost that the watchcases would be dust-tight and robust. In 1926, he patented his "Oyster" case. This consisted of all the elements for the middle section of the case, all screwed together with the case back and the bezel with the crystal and with the winding crown. Thus, the entire case was extremely stable.

Compared to other dustproof and waterproof watchcases from the 1920s, the "Oyster" from Rolex was flat and compact and could also be worn on slender forearms, inspiring Hans Wilsdorf to come up with a bold marketing idea. He provided Mercedes Gleitze—an "extreme swimmer," as we would say today—with an Oyster for her attempt to swim across the English Channel, and when this athletic stenographer and typist actually managed to swim the Channel on her second attempt, the news was emblazoned in large type on the front page of the *London Daily Mail* the next day: Rolex "Defies the Elements." London concessionaires displayed their Oysters in a water-filled goldfish bowl . . .

The Oyster case with its screw-down bezel, crown, and case back has guaranteed its water resistance since 1926. The expedition into the Mariana Trench required a thicker crystal.

4 The Perpetual Rotor: ALWAYS IN MOTION

The first central rotor from 1931 wound up in only one direction. Later, red anodized aluminum wheels became the symbols of bidirectional winding.

Twenty-three hours and fifty-nine minutes a day, the precision watch movements were housed, dry and safe, in the all-around screw-down Oyster case. But the wearer had to disturb this hermetic idyll on a regular basis and unscrew the crown to give the movement new energy. For a perfectionist such as Hans Wilsdorf, this violation of the protective atmosphere must have been very painful, but the methods for automatically winding a watch movement available up to that time—using pendulum weights or movable lugs—were not satisfactory for him. It was not until 1931, when the Aegler watch movement factory developed a centrally mounted, rotating oscillating weight or rotor—which rotated freely in one direction and retensioned the mainspring in small increments in the other direction with significant reduction—that Wilsdorf had his "perpetual" self-winding mechanism. Due to its compact design, the Perpetual rotor could fit under the bulbous screw back of the Oyster case, and from then on, no Rolex owner had to be concerned about winding it any longer—all they had to do was wear the watch.

In 1933, Rolex was granted a patent for the rotor winding mechanism, and for fifteen years the competition had to make do with "automatic hammer" mechanisms and pendulum oscillating weights that were clearly inferior to the Rolex perpetual rotor in terms of how they operated and their efficiency. Only after the Rolex patent expired in 1948 was it possible for the automatic self-winding mechanism to start its triumphal march on a broad front.

In 1959, Rolex developed a robust, reliable, and compact ratchet reversing wheel that uses the rotor's movements in both directions to wind the barrel. The two red anodized reversing wheels made of duralumin are still a distinctive feature of this design today.

Displays and Functions:
THE DATE—JUST IN TIME

The watch world owes a lot to Rolex because it was this watch manufacturer that first actually introduced many of the functions and displays we take for granted on a wristwatch today.

For example, the date window at the "3": first seen in 1945 on a gold model of the Oyster Perpetual with the evocative name of Datejust, because the jumping changeover was made at exactly midnight. Beginning with this model, the Oyster collection has developed in two directions. On one hand, Rolex created more and more elegant calendar watches, such as the Day-Date from 1956, which displayed the date and day of the week for the first time; the Sky-Dweller, presented in 2012, with an annual calendar and a second time zone, also falls into this category. On the other hand, Rolex developed the so-called tool watches, which the manufacturer calls the "Professionals." On these models everything comes down to good readability, unconditional robustness, and useful functions. Mention should be made here of the Submariner, presented in 1953, which was one of the first diving watches to be equipped with a rotating bezel with sixty graduations.

The additional twenty-four-hour hand for displaying a second zone time made its debut in the GMT-Master in 1954. Its successor model, the GMT-Master II, for the first time featured an incrementally adjustable main hour hand to quickly adjust the display to a new local time, while the twenty-four-hour hand (with an arrowhead) remained set to the time at home.

The ten-minute regatta countdown indicator on the Yacht-Master II.

The Sky-Dweller has a second zone time display with a twenty-four-hour disc, as well as an annual calendar, with the month displayed via the color of the window above the hour marker.

The characteristic position of the date and day of the week is a distinguishing feature of the Day-Date.

An orange-colored, arrow-shaped hand for the second zone time on the Explorer II

To better highlight the date display, Rolex grinds a magnifier into the sapphire crystal.

6 Hairsprings of Metal and Silicon: OSCILLATORY CHARACTERISTICS

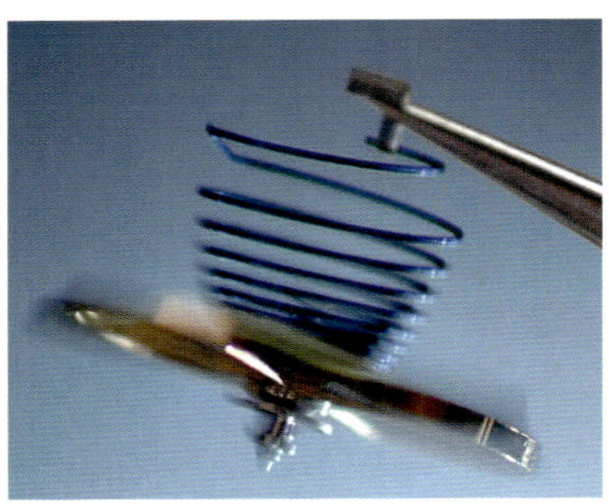

Rolex is one of the few watch manufacturers to make its own balance wheels. Due to the effective division of labor in the Swiss watch industry, for a long time there had been no reason for a watch brand to become involved in the highly specialized production of escapement parts. It was only when the quasi-monopoly Nivarox (the Swatch Group) began to think aloud about regimenting supplies that some of the larger luxury manufacturers invested in making their own hairsprings.

In 2000, Rolex developed the Parachrom hairspring, a conservative metal spring made of a niobium-zirconium alloy with good antimagnetic properties and stability at high temperatures. The Parachrom hairspring gets its characteristic blue color from a surface treatment introduced in 2005 to improve its long-term stability.

While the Parachrom hairspring is used in the larger men's watch movements, the Syloxi hairspring, presented in 2014 in the caliber 2236, is more suitable for smaller watches. It is made of a crystalline composite material of silicon and silicon oxide (hence the name Syloxi); this has temperature-equalizing and antimagnetic properties that give it very high resistance to interference. Besides this, Rolex has optimized the oscillation behavior by means of a patented geometry it has developed: The variable pitch and thickness of the coils ensure the constant "breathing" of the hairspring in any position.

The Parachrom hairspring is made of a metal alloy, while in contrast, the Syloxi hairspring shown below is made of polycrystalline silicon.

7 The Chronergy Escapement: GOOD IS GOOD, BUT BETTER CARRIES IT

Caliber 3255, presented in 2015, established a new philosophy of design. The great challenges for the architects of the new watch era include a high level of efficiency, a large power reserve, absolute precision, and maximal insensitivity to magnetism. Rolex engineers did the only right thing and started their work on a blank sheet of paper.

The heart of the new design is the escapement, patented under the name Chronergy, which in fact is based on the principle of the Swiss lever escapement. However, it has a 15 percent higher degree of efficiency thanks to its new geometry and consistent lightweight design. The gain in impulse power is actually significantly greater, because the conventional lever escapement transmits only about one-third the energy supplied by the barrel to the balance wheel and hairspring. As a result, this escapement is responsible for almost half the increase in the movement's power reserve (to seventy hours). A new barrel with a longer spring contributes the other half.

The anchor and escapement wheel—that is, the delicate key components of the Chronergy escapement—are manufactured from a nickel-phosphorus alloy using the LIGA process (electrogalvanic molding) and are thus insensitive to magnetic fields. The ruby pallet jewels of the anchor measure 0.125 mm and are only half as wide or heavy as those of the previous generation, while the contact area of the escapement wheel teeth has been doubled.

Improved efficiency and lower weight characterize the Chronergy escapement, which nevertheless remains true to the principles of the Swiss lever escapement.

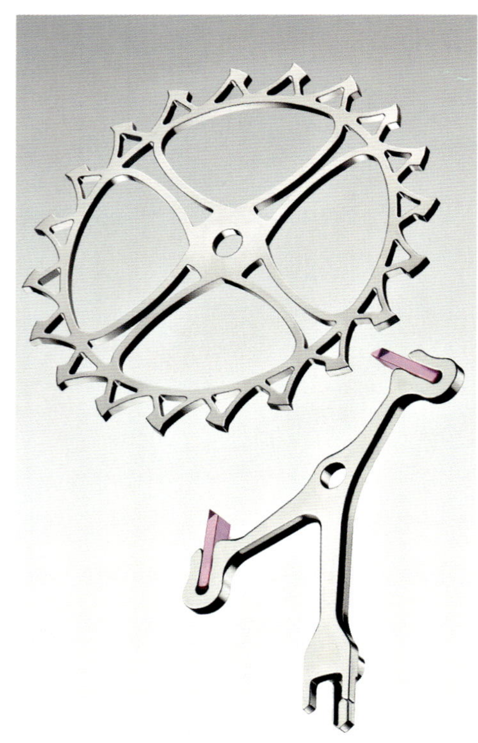

8 Chromalight Luminous Phosphorescent Material: LIGHT AT THE END OF THE TUNNEL

For some years now, on all the sports and many of the classic Oyster models, the hands, hour markers, and other display elements have emerged in bright blue as soon as the watch is plunged into darkness. The Chromalight luminous material, used exclusively by Rolex to coat or fill in these various components, is particularly efficient and significantly surpasses the standards prevalent in the watchmaking industry in terms of duration and intensity. The Chromalight display luminescence lasts practically twice as long as a standard luminous material does. Besides this, the luminosity itself works in a way that is distributed more evenly over the entire duration of the luminous time, which can be more than eight hours.

The basic material used is an ultrafine oxide powder—with aluminum, strontium, dysprosium, and europium compounds—which, when fired at high temperatures, forms crystals whose specific atomic structure produces the emission of the typical blue light. After this, the material has phosphorescent properties; that is, the ability to store light energy and gradually release it again in the dark. The phenomenon of light emission is based on the movement of electrical charges within matter. The phosphorescent powder is mixed with liquid resin and applied by hand to the dials, hour markers, and hands.

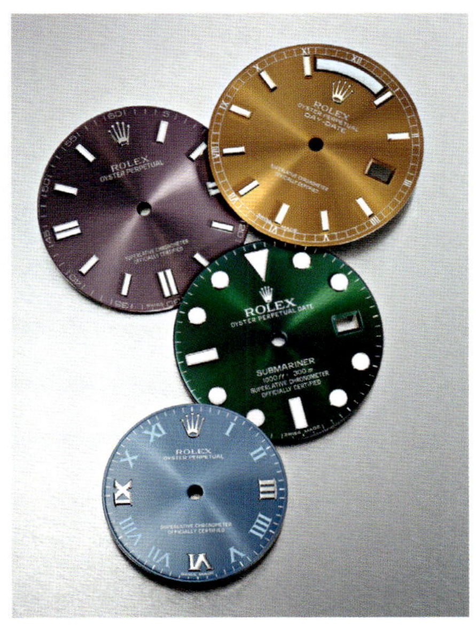

The Chromalight elements on the dial of the Deepsea Challenge (*above*) shimmer in blue light. The hands are also coated with the pasty, luminous paint.

The Cerachrom Bezel:
HARD, HARDER, CERAMIC

The fields of color on the Cerachrom bezel appear separated by a hair's breadth.

On a wristwatch, the bezel is not only one of the most prominent design elements, but also one of the components most exposed to stresses, such as bumps, scratches, corrosion, and other environmental impacts. A Cerachrom number disc was introduced on the GMT-Master II in 2005. By 2013, the new Cosmograph Daytona had a Monobloc bezel made entirely of Cerachrom.

The basic material is a very fine zirconium or aluminum oxide powder made of particles that are less than a thousandth of a millimeter in diameter. A binder and pigments are added to the powder to create the desired color. The material, mixed in liquid form, is injected under high pressure into a mold, creating a blank for the component. In this work step the component is given its geometric shape, and all the digits, graduations, and inscriptions are created—either recessed or raised. After demolding and "debinding" (over several days at low heat), the blank is sintered at approximately 2,900°F and thus becomes harder than steel. During the hardening process, the component shrinks in a calculable and symmetrical manner by about 25 to 30 percent and assumes its final color shade, or shades. In the last processing step, diamond milling is used to trim each part to its exact dimensions.

Recessed digits, graduations, and inscriptions are plated with a thin layer of yellow gold, rose gold, or platinum by using the PVD process. During the final polishing process, the metal layer is removed from the surface of the number disc, and the inscriptions are displayed to their best advantage by the precious metal that remains in the recesses.

The process for manufacturing two-toned Cerachrom number discs in the form they are used on the GMT-Master II consists of converting part of the initially single-color disc. For the blue-black number disc the blue is transformed into black, while on the red-blue number disc the red is changed into blue. To do this, half of the number disc is impregnated with an aqueous solution before sintering. During this heat treatment the ceramic thickens and the added compounds react with the basic substance of the Cerachrom number disc. This subsequently changes the original color shade in one-half of the disc.

10 Bracelets and Clasps: OYSTERS, JUBILEES, AND PRESIDENTS

The quality of Rolex link bracelets has noticeably improved in the last ten years—not that the stainless-steel and gold bracelets had previously ever attracted attention for poor workmanship or functional deficits. But in terms of manufacturing tolerances, a lot has happened in the third millennium, and customer demands for wearing comfort have also increased significantly. Rattling is no longer tolerated.

In the Rolex nomenclature, you distinguish between the three-piece link Oyster bracelet, which has been in use since the 1930s and is part of the basic equipment of most "tool watches," and the five-piece-link Jubilee bracelet, which was introduced in 1945 for the launch of the Datejust. The refined President bracelet, with its rounded elements, was introduced along with the Day-Date in 1956 and is still reserved for this model and some other versions of the Datejust. It is available only in a precious metal and is equipped with a concealed crown clasp. The five-piece Pearlmaster link bracelet (since 1992) is designed in a very similar way.

A distinct folding clasp was developed for each bracelet; these all offer a secure closure and meet the highest requirements in terms of wearing comfort and ergonomics. Their reliability is tested by opening and closing them tens of thousands of times.

The Oysterclasp, with its foldout extension links, has a two-part design; its front edge acts like a lever and makes opening it easier. The Oysterlock clasp has an additional safety catch that prevents the wearer from opening it unintentionally. In the latest version, the Crownclasp buckle is discreetly hidden under a small lever in the shape of the Rolex crown. It comes on the President and Pearlmaster bracelets, as well as on certain of the Jubilee bracelets. The Glidelock fine adjustment system allows the wearer to extend the bracelet in fine increments by up to 20 mm via a toothed rack fastened to the cover of the clasp. This is used on the Submariner, Sea-Dweller, and Deepsea diver watch models.

Peter Braun

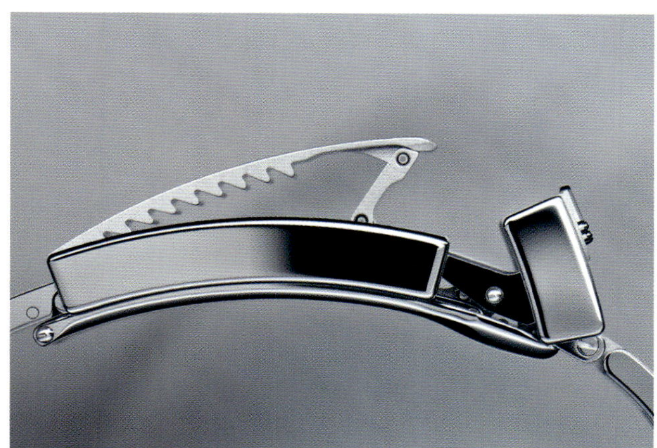

Fine adjustment integrated in the clasp: the Glidelock

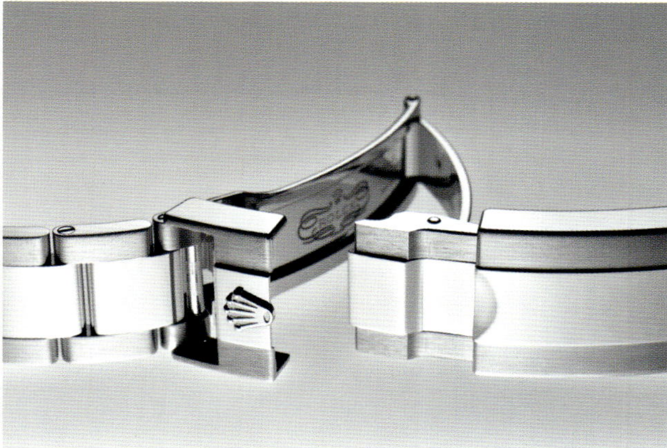

The Oysterclasp has a safety cover and a fold-over extension link.

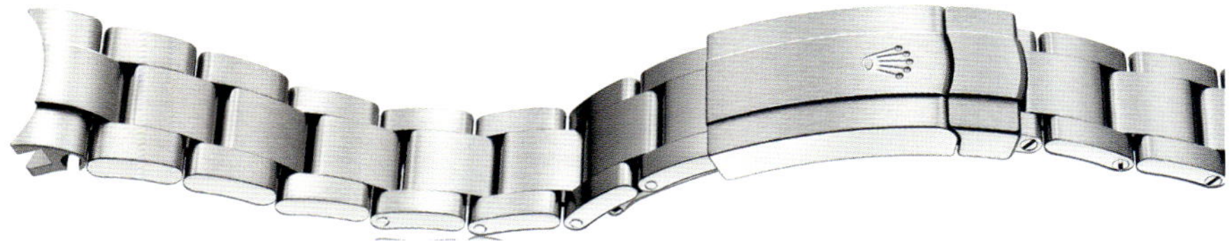

Classic three-piece link design: the Rolex Oyster link bracelet

The elegant Pearlmaster models have a bracelet design of their own.

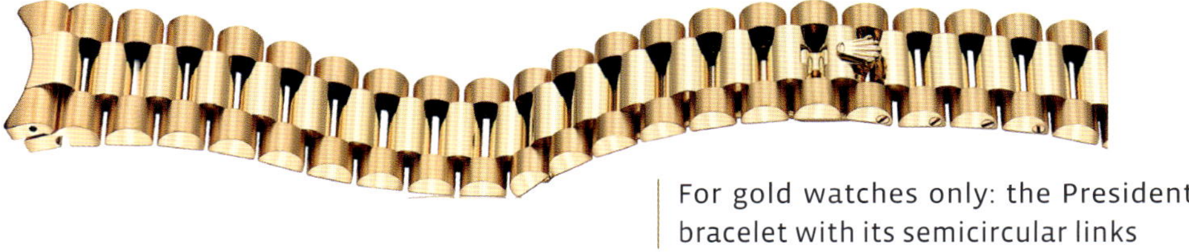

For gold watches only: the President bracelet with its semicircular links

GENEVA
ROLEX
POS. + TEMP.

VII.
The New Generation of Calibers

A New Era in Reckoning Time

A 15 percent higher escapement efficiency and 50 percent more power reserve, with 90 percent newly developed parts and an astonishing 100 percent higher level of precision than the COSC chronometer test requires: with the caliber 3255, a new era in reckoning time began for the company in 2015.

The new caliber 3200 family has meanwhile become the standard equipment of the Oyster Perpetual model series. Pictured is the caliber 3255 with simultaneous date and day-of-the-week display.

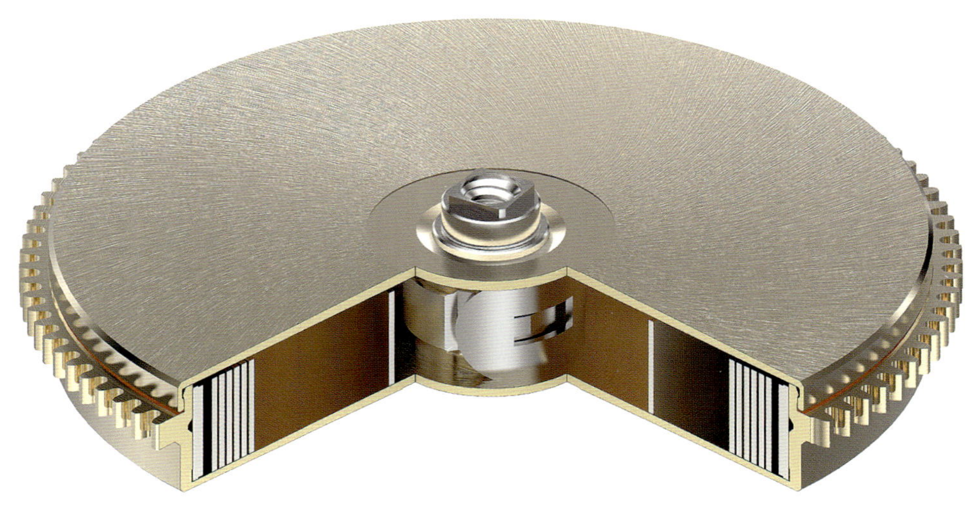

The first Rolex model to be fitted with the caliber 3255 was the Oyster Perpetual Day-Date 40 in 2015.

The particularly thin-walled barrel offers space for a longer mainspring. This enables the power reserve to increase to almost three days.

For more than a hundred years, their precise, reliable, low-maintenance, and durable self-winding movements have been playing a significant role in the good reputation of the wristwatches with a crown in the signet. As early as the 1920s, Rolex proved that it was possible to achieve chronometer rates even with small watch movements, and thanks to the waterproof, screw-down case back and crown tube, this precision was soon no longer affected by any external interferences. When, in 1931, the bidirectional automatic rotor winding system with a simple change gear mechanism also showed it could function smoothly, the prototype of the modern wristwatch had been created.

From then on, the Rolex Oyster was refined only in terms of details—and the same is true for the watch movement, which was produced by the autonomous Biel-based company Aegler. This had, in fact, already been renamed Rolex Manufacture a long time ago, but the factory did not become an integral part of the Rolex company until 2004, shortly before the one-hundredth birthday of the brand, which had long become world famous.

NEW DEVELOPMENTS

The caliber 3255, presented at Baselworld 2015, is the first movement of the united Rolex company, and, after the chronograph caliber 4130, dating from 2000, it is the first completely redesigned self-winding movement from the Rolex brand. Sooner or later the entire caliber 3200 family will gradually replace the caliber 3100 movements. These, despite constant improvements, are slowly getting on in years.

The great challenges facing the architects of the new watch era are a high level of efficiency, a large power reserve, absolute precision, and maximum insensitivity to magnetism. Rolex engineers did the only right thing and started their work on a blank sheet of paper.

The movement diameter was increased slightly (from 28.5 to 29.1 mm), and it was possible to reduce the overall height from 6 to a little over 5 mm by saving space in the gearbox and by doing without an additional heavy metal cheek on the winding rotor.

There are fourteen patents protecting these innovative design details, which in total not only increase the power reserve but also improve precision, thanks to the more efficient escapement and the friction-optimized wheel train bridge.

THE CHRONERGY ESCAPEMENT

The heart of the new design is the escapement patented under the name Chronergy, which is based on the principle of the Swiss lever escapement, but thanks to its new geometry and systematic lightweight construction, it is 15 percent more efficient. The gain in impulse power is significantly greater because the conventional lever escapement transmits only about one-third of the energy supplied by the barrel to the balance wheel and hairspring. As a result, this escapement is responsible for almost half of the increase in the movement's power reserve. The other half is contributed by a high-precision barrel with around 50 percent less wall thickness. This provides space for a longer spring, which additionally stores more than ten hours of power reserve.

The anchor and escapement wheel—the delicate key components of the Chronergy escapement—are made of a nickel-phosphorus alloy using the LIGA process (electrogalvanic molding) and are therefore insensitive to magnetic fields. The ruby pallet jewels of the anchor measure 0.125 mm and are only half as wide or heavy as those of the previous generation, while the contact area of the escapement wheel teeth has been doubled.

The caliber 3255 balance wheel has a blue Parachrom hairspring manufactured by Rolex and made from a patented niobium-zirconium alloy. The nonmagnetic hairspring maintains temperature stability and is reported to be up to ten times less sensitive to shocks than a conventional hairspring. It is provided with an optimized Breguet overcoil hairspring, ensuring better isochronism of its oscillations in all positions.

The large balance wheel with variable inertia has four Microstella nuts for fine adjustment. Thanks to the revised geometric shape and more-precise machining, their sphere of action has been improved by a factor of three.

The oscillating system (balance wheel and hairspring) is held in place with a Paraflex shock absorber on a traversing balance bridge that has an optimized height adjustment system.

OPTIMIZED WHEEL TRAIN BRIDGE AND AUTOMATIC FUNCTION

The wheel train bridge efficiency has been optimized by means of newly shaped teeth and new pairings of materials. In addition, Rolex has also developed new lubricants that are manufactured as synthetics in-house. These are reported to feature longer service life and better long-term durability. The caliber 3255 has a new-generation winding mechanism that can wind up the new high-performance barrel faster. The reversing system has been optimized with a view to higher efficiency. The Monobloc oscillating-weight rotor, now made from one piece, has ball bearings and is held by a central screw.

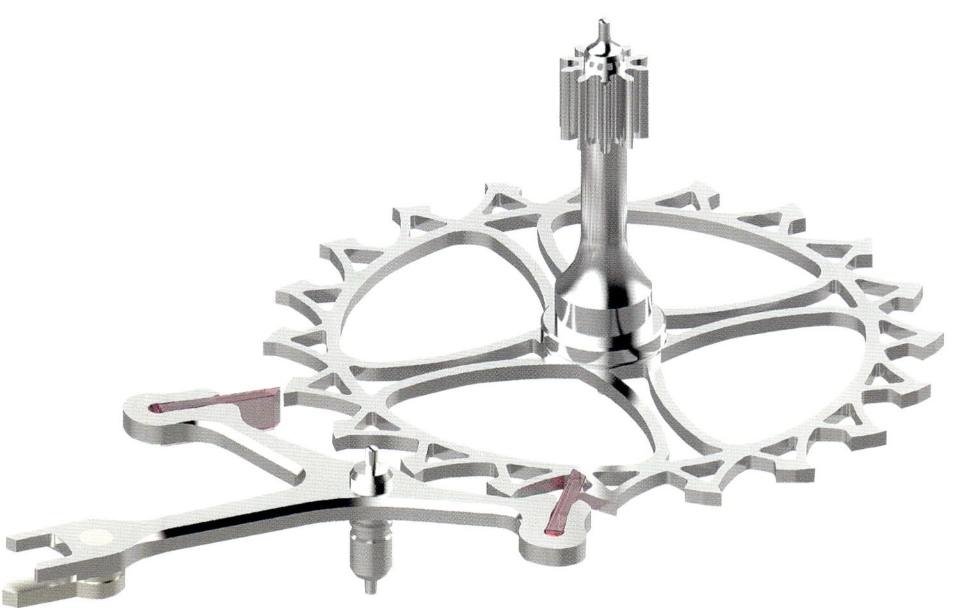

Despite the many new design features and its modern architecture, the caliber 3255 has all the features of a typical Rolex movement. Of course, the red reversing wheels in the winding mechanism are not to be missed, although these days they are no longer made of anodized duralumin.

With the first movement of the new generation, Rolex is setting a new standard in the field of precision that goes beyond the criteria of the Official Swiss Chronometer Testing Institute (COSC). The brand has developed new high-tech processes and equipment to test the accuracy of its "superlative chronometers," with tolerances twice as strict as those of the COSC (+2/–2 seconds per day instead of +6/–4 seconds), doing this under conditions that simulate a real-life situation more appropriate to the wearer's everyday life. Rolex has developed a specific test protocol based on large-scale statistical research to determine the real-life conditions of wearing a wristwatch. These exclusive tests complement official certification by the COSC, to which all Rolex movements are still systematically subjected. However, they are extended to the entire fully assembled watch after the movement has been encased.

Peter Braun

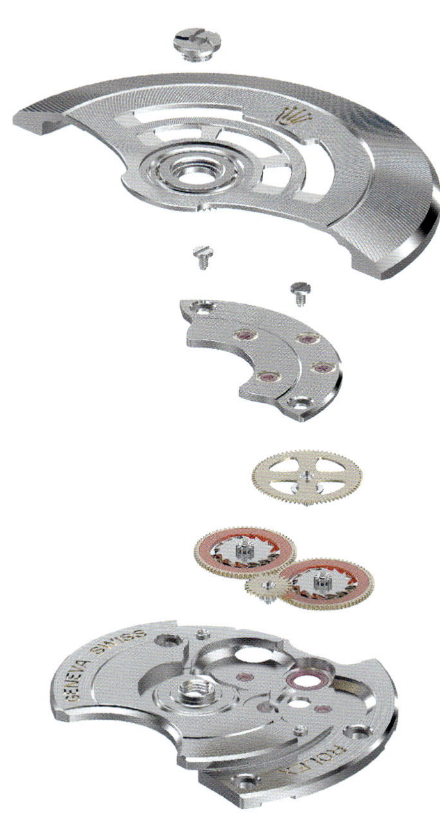

The anchor and escapement wheel of the Chronergy escapement have a very narrow design. The LIGA production process makes it possible to make delicate structures and ensure a low dead weight.

The rotor winding mechanism still works in both directions of rotation; the rotor is now made of only one piece.

The Current Self-Winding Movements

The Rolex caliber range is changing radically. The successful new developments in the 3200 series are putting the designers on the spot. Further steps in development can be expected for the other caliber families in the coming years.

With the caliber 3255, a new era of reckoning time began for the company in 2015. The new developments to replace the veteran caliber 3100 family are impressive, with a 15 percent higher efficiency in the escapement, 50 percent more power reserve, 90 percent newly developed parts, and an astonishing 100 percent higher level of precision than is required by the COSC chronometer test.

A year earlier, in 2014, the caliber 9001, with a technically brilliant annual calendar and second time zone in twenty-four-hour format, had been introduced, although—like the chronograph movement caliber 4130 (and its derivative 4161, with a "backward" counter)—it is still based on traditional Rolex watch movement technology.

Traditional here means doing without new synthetic materials, without the use of special tooth profiles and without a lightweight barrel. The Glucydur balance wheel, which has Microstella screws to adjust the mass inertia and a blue Parachrom hairspring, has been established for several years now and undoubtedly contributes its part to the very good rate values of the Rolex in-house watch movements.

INCREASING REQUIREMENTS

Fourteen patents protect the innovative design details, which overall both increase the power reserve as well as improve precision due to the more efficient escapement and the friction-optimized wheel train bridge. With the new caliber 3235–3255 generation, Rolex is setting a new standard in the field of rate precision that goes beyond the criteria of the Official Swiss Chronometer Testing Institute (COSC). The brand developed new high-tech procedures and equipment to test the accuracy of its "superlative chronometers," with tolerances twice as strict as those of the COSC (+2/ –2 seconds per day instead of +6/–4 seconds) and under conditions that simulated a real-life situation more in line with everyday life. Rolex has developed a specific test protocol based on large-scale statistical studies to determine the real conditions for wearing a wristwatch. These exclusive tests complement official certification by the COSC, to which all Rolex movements are still systematically subjected. However, they extend to the entire, fully assembled watch after the movement has been encased. On the following pages we provide you with an overview of the calibers currently in use that can be assigned to different generations. Calibers 3135 and 3186 are still based on the old architecture; caliber 9001, like the two chronographs (4130 and 4161), is in the new twenty-first century designs. Caliber 2236 has already been "modernized," and the caliber twins 3235–3255 point the way to the future.

Caliber 3235

Self-winding; optimized Chronergy escapement, anchor, and escapement wheel made of nickel-phosphorus using the LIGA process; simple barrel, power reserve 70 hours; certified chronometer (COSC)

Functions: Hours, minutes, center second; date (jumping)
Diameter: 29.1 mm
Height: 5.4 mm
Jewels: 31
Balance wheel: Glucydur with Microstella nuts
Vibrations per hour: 28,800 vph
Hairspring: Parachrom Breguet hairspring
Shock absorber: Paraflex
Remarks: Used in such models as the Oyster Perpetual Datejust 41

Caliber 3255

Self-winding; optimized Chronergy escapement, anchor and escapement wheel made of nickel-phosphorus using the LIGA process; simple barrel, power reserve 70 hours; certified chronometer (COSC)

Functions: Hours, minutes, center seconds; date and day of the week (both jumping)
Diameter: 29.1 mm
Height: 5.4 mm
Jewels: 31
Balance wheel: Glucydur with Microstella regulating nuts
Vibrations per hour: 28,800 vph
Hairspring: Syloxi flat hairspring
Shock absorber: Paraflex
Remarks: Used in the Oyster Perpetual Day-Date 4 model

Caliber 2236

Self-winding; simple barrel, power reserve 55 hours; certified chronometer (COSC)

Functions: Hours, minutes, center seconds; date (jumping)
Diameter: 20 mm
Height: 5.95 mm
Jewels: 31
Balance wheel: Glucydur with Microstella regulating nuts
Vibrations per hour: 28,800 vph
Hairspring: Syloxi flat hairspring
Shock absorber: Paraflex
Remarks: Used in the Yachtmaster 37 (image) and Lady Datejust models

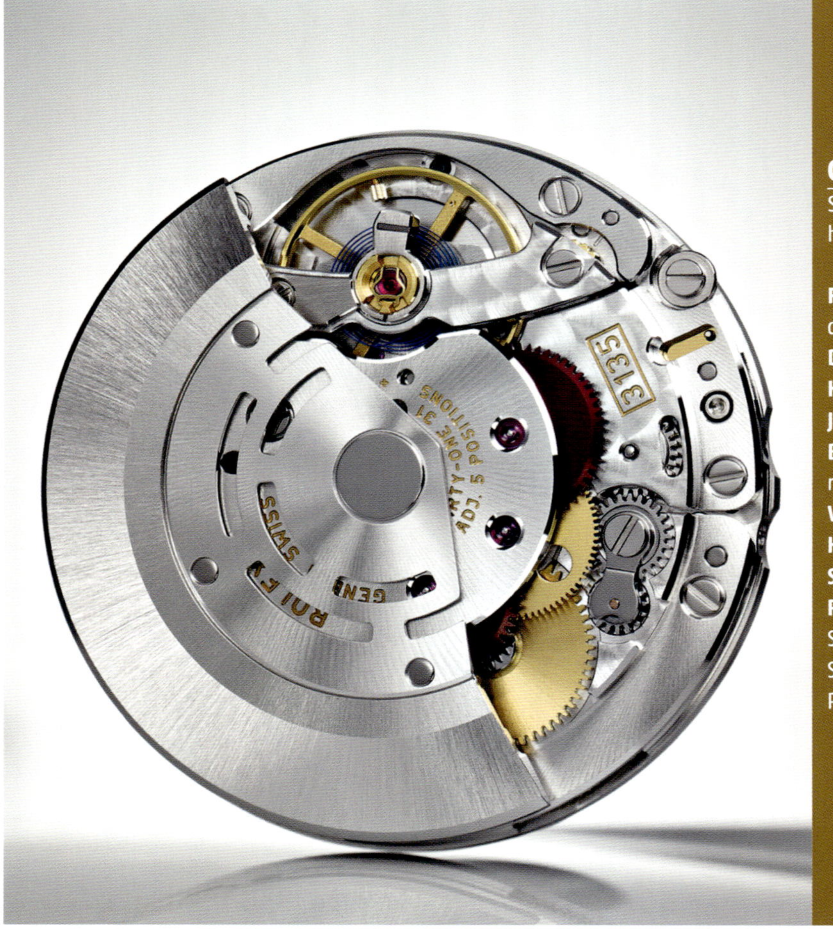

Caliber 3135

Self-winding; simple barrel, power reserve 48 hours; certified chronometer (COSC)

Functions: Hours, minutes, center seconds; date (jumping)
Diameter: 28.5 mm
Height: 5.95 mm
Jewels: 31
Balance wheel: Glucydur with Microstella regulating nuts
Vibrations per hour: 28,800 vph
Hairspring: Parachrom flat hairspring
Shock absorber: KIF
Remarks: Used in the Yachtmaster 40 (image), Submariner Date (until 2020), Sea-Dweller 4000, Sea-Dweller Deepsea, Datejust 36, and Oyster Perpetual Date models

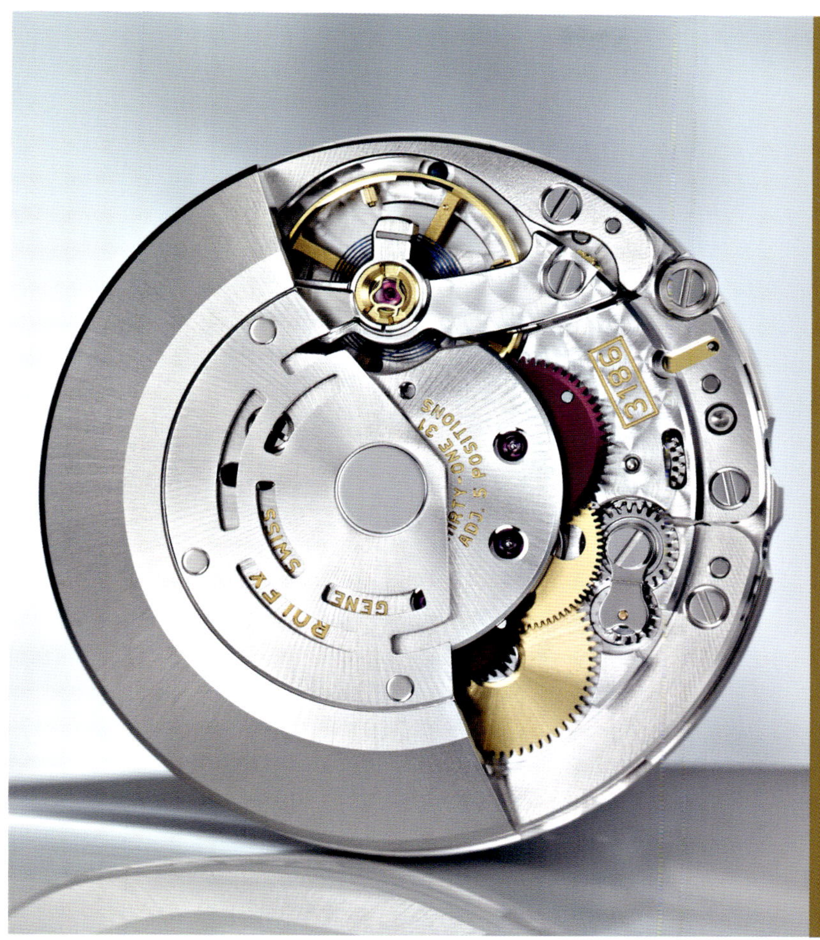

Caliber 3186

Self-winding; simple barrel, power reserve 48 hours; certified chronometer (COSC)

Functions: Hours (adjustable by the crown in hourly increments), minutes, center seconds; 24-hour pointer; date (jumping)
Diameter: 28.5 mm
Height: 6.05 mm
Jewels: 31
Balance wheel: Glucydur with Microstella regulating nuts
Vibrations per hour: 28,800 vph
Hairspring: Parachrom flat hairspring
Shock absorber: KIF
Remarks: Used in the GMT-Master II (image) and Explorer II (caliber 3187) models

Caliber 9001

Self-winding; simple barrel, power reserve 72 hours; certified chronometer (COSC)

Functions: Hours, minutes, center seconds; additional 24-hour display (second time zone); annual calendar with date, month
Diameter: 33 mm
Height: 7.2 mm
Jewels: 40
Balance wheel: Glucydur with Microstella regulating nuts
Vibrations per hour: 28,800 vph
Hairspring: Parachrom Breguet hairspring
Shock absorber: Paraflex
Remarks: Used in the Sky-Dweller model

Caliber 4130

Self-winding; simple barrel, power reserve 72 hours; certified chronometer (COSC)

Functions: Hours, minutes, small seconds; chronograph
Diameter: 30.5 mm
Height: 7.5 mm
Jewels: 44
Balance wheel: Glucydur with Microstella regulating nuts
Vibrations per hour: 28,800 vph
Hairspring: Parachrom Breguet hairspring
Shock absorber: KIF
Remarks: Used in the Cosmograph Daytona model

Caliber 4161

Self-winding; simple barrel, power reserve 72 hours; certified chronometer (COSC)

Basic caliber: 4130
Functions: Hours, minutes, small seconds; programmable regatta countdown with memory
Diameter: 31.2 mm
Height: 8.05 mm
Jewels: 42
Balance wheel: Glucydur with Microstella regulating nuts
Vibrations per hour: 28,800 vph
Hairspring: Parachrom Breguet hairspring
Shock absorber: KIF
Remarks: Used in the Yacht-Master II model

THE NEW ROLEX WATCH MOVEMENT TECHNOLOGY

SELF-WINDING MOVEMENT

As early as the 1920s, Rolex—or rather, its movement manufacturer at the time, Aegler SA in Biel, which would later become Rolex Manufacture SA—proved that even small watch movements could achieve chronometer rates. Thanks to the watertight, screw-down case back and the crown tube, this precision would soon no longer be affected by external interferences. The so-called Oyster case made it possible to create the first really practical and "wearable" waterproof wristwatch, because the solutions previously offered by various other companies were neither attractive nor easy to work with.

The big breakthrough for the wristwatch came in 1931, with the automatic rotor winding mechanism, which always ensured an energy-charged mainspring and relieved the wearer of having to devote any thought to making sure their watch would run. Basically, the wristwatch has not evolved very much in the last eighty-five years.

Pendulum oscillating weights with buffer springs to limit how far the weight travels and other inventive mechanisms used by various watch movement manufacturers during the 1940s disappeared from the market almost instantly when the Rolex patent on the "Perpetual rotor" expired in 1948.

In 1959, Rolex engineers succeeded in vastly improving the efficiency of the rotor winding mechanism. By inserting a synchronizer control gear to switch the direction of rotation, they were able to harness the rotor's motion in both directions to wind the barrel. The two red anodized aluminum reversing wheels have been a distinctive identifying feature of Rolex self-winding movements ever since.

All the winding rotors and the winding mechanism components are manufactured and preassembled in the Rolex watch factory in Biel. However, they are put in place in the movements only during final assembly at Rolex's production location at Les Acacias, Geneva. The new calibers 3235 and 3255 are an exception; on these, the rotors, milled from a single piece, are already mounted in Biel.

The new generation of winding mechanisms can wind up the new high-performance barrel faster. The reversing system has been optimized in view of higher efficiency.

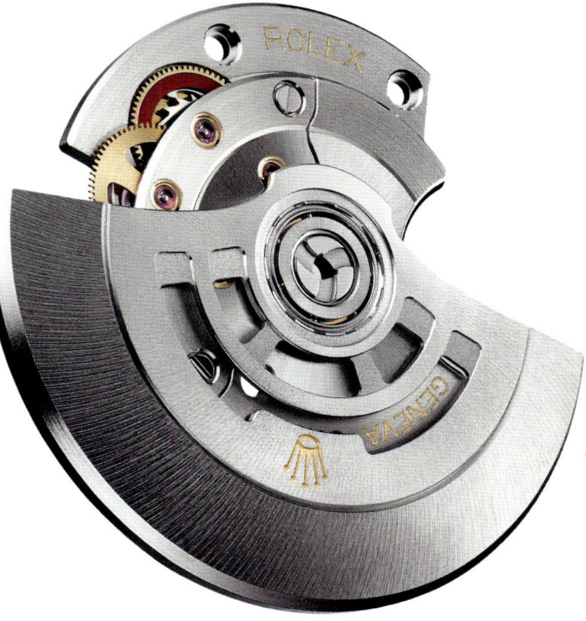

The new rotors (here caliber 3255) are made from one piece. On the others, the heavy metal cheek is riveted to the rotor.

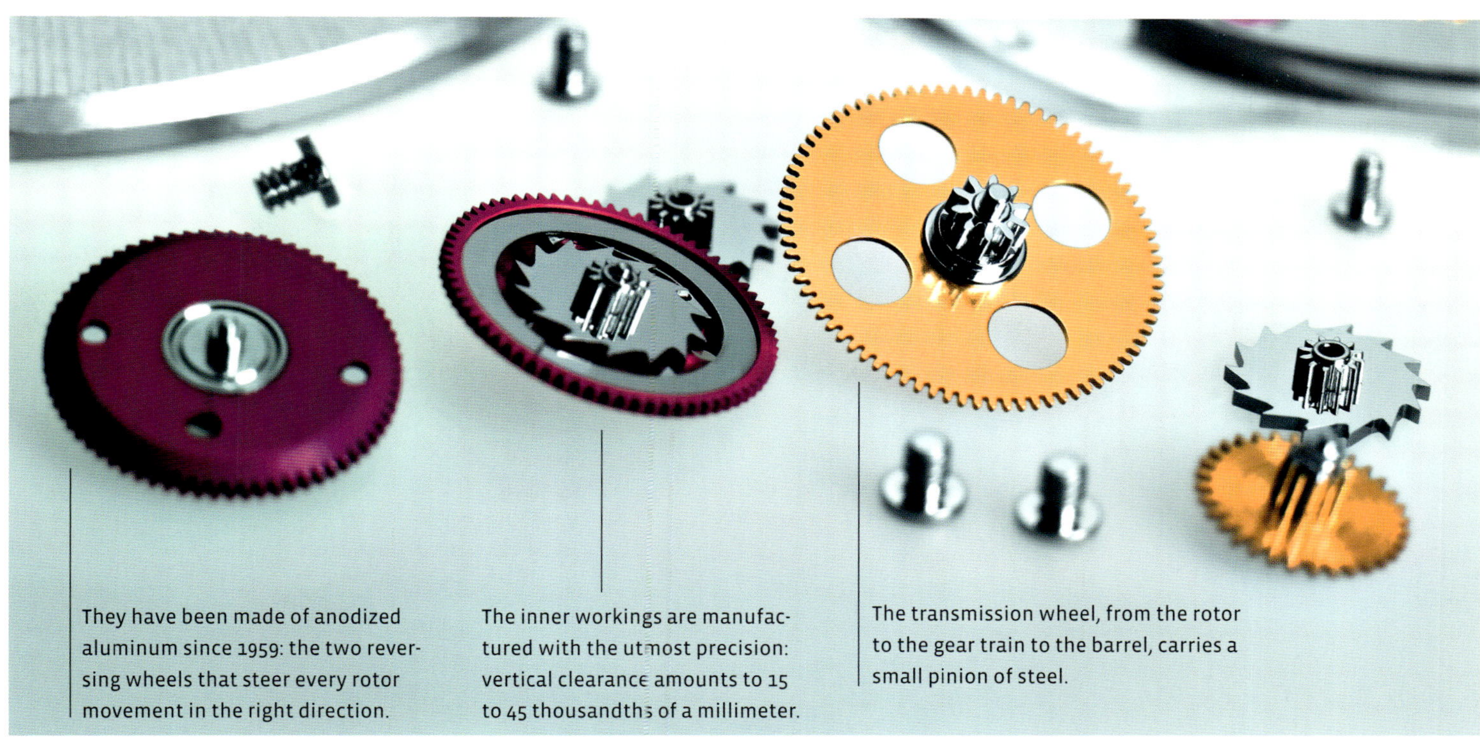

They have been made of anodized aluminum since 1959: the two reversing wheels that steer every rotor movement in the right direction.

The inner workings are manufactured with the utmost precision: vertical clearance amounts to 15 to 45 thousandths of a millimeter.

The transmission wheel, from the rotor to the gear train to the barrel, carries a small pinion of steel.

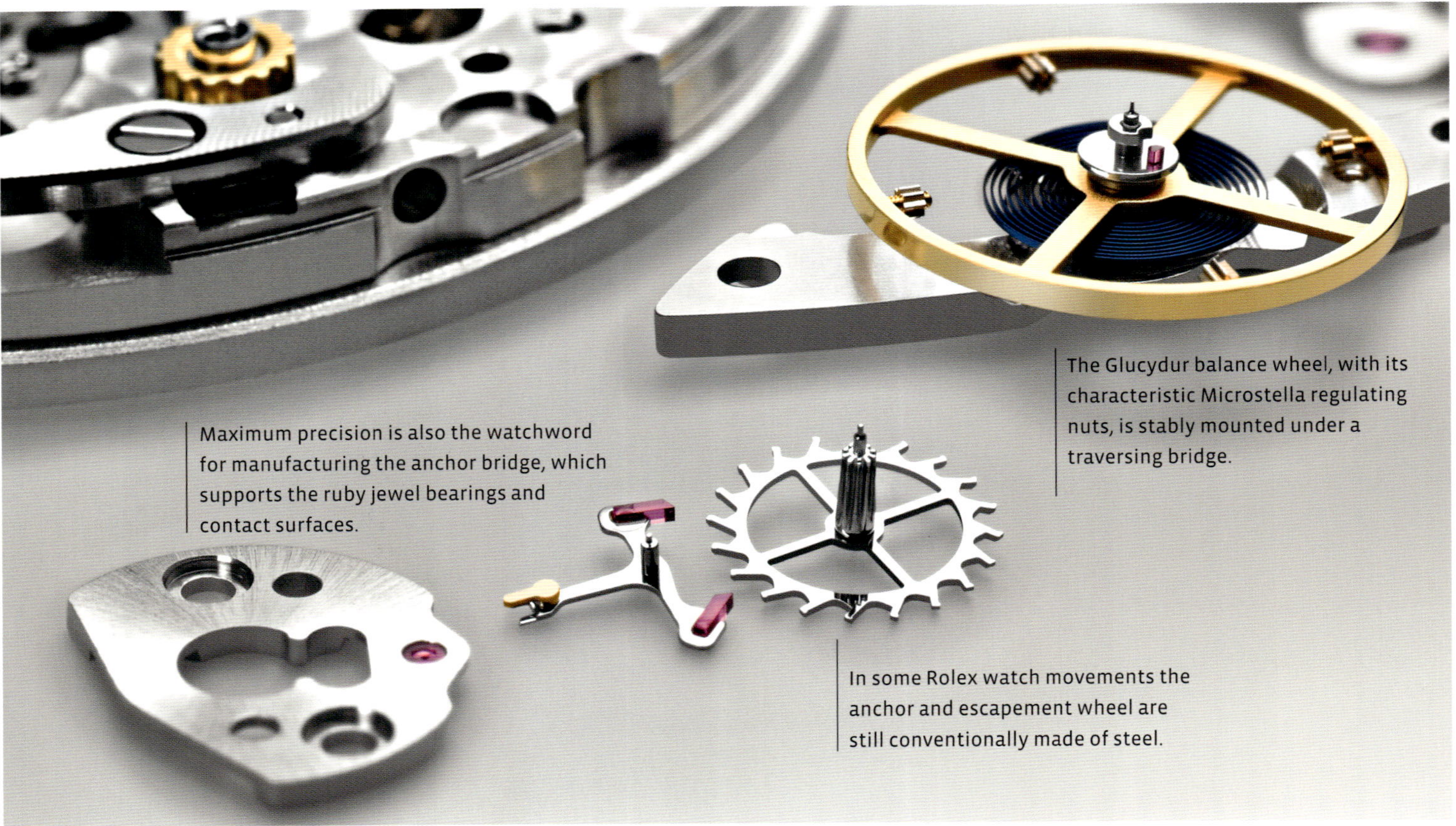

Maximum precision is also the watchword for manufacturing the anchor bridge, which supports the ruby jewel bearings and contact surfaces.

The Glucydur balance wheel, with its characteristic Microstella regulating nuts, is stably mounted under a traversing bridge.

In some Rolex watch movements the anchor and escapement wheel are still conventionally made of steel.

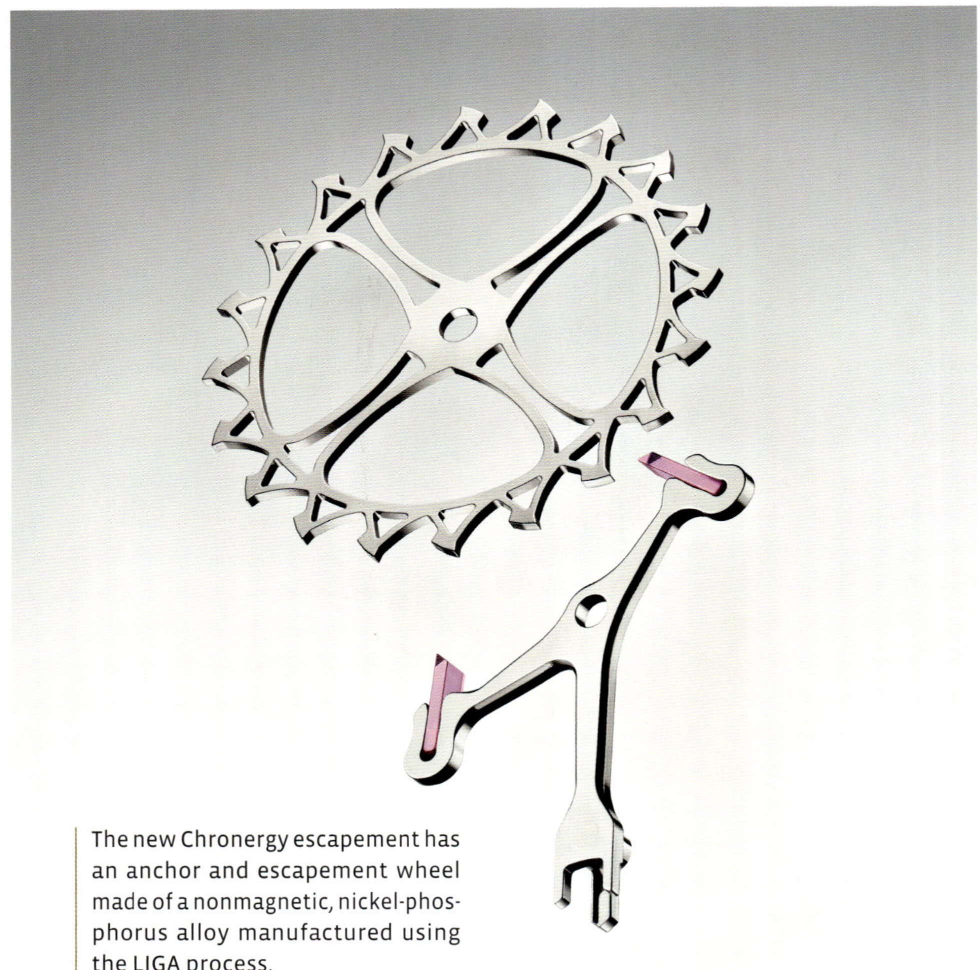

The new Chronergy escapement has an anchor and escapement wheel made of a nonmagnetic, nickel-phosphorus alloy manufactured using the LIGA process.

THE CHRONERGY ESCAPEMENT

The rotor isn't the only exceptional component on the two new calibers; these differ from the last generation of calibers in almost every detail and adopted at most 10 percent of the individual parts without any changes. The heart of the new design is the escapement, patented under the name Chronergy, which in fact is based on the principle of the Swiss lever escapement. Thanks to its new geometry and systematic lightweight design, it is 15 percent more efficient. Yet, the gain in impulse power is significantly greater than that because the conventional lever escapement transmits only about one-third the energy supplied by the barrel to the balance wheel and hairspring. As a result, the escapement is responsible for almost half the increase in the watch movement's power reserve. The other half is supplied by a high-precision barrel, with its wall thickness reduced by around 50 percent: It offers space for a longer mainspring that in addition stores more than ten hours of power reserve.

The anchor and escapement wheel, the delicate key components of the new Chronergy escapement, are made of a nickel-phosphorus alloy using the LIGA process (electrogalvanic molding) and are therefore insensitive to magnetic fields. The anchor's ruby pallet jewels measure 0.125 millimeters and are only half as wide or heavy as those of the previous generation, while the contact area of the escapement wheel teeth has been doubled.

The efficiency of the wheel train bridge has been optimized by the new shape of the teeth and new material pairings. Besides this, Rolex has developed new lubricants that are synthetically manufactured in-house. These are reported to feature a longer service life and higher long-term stability.

NEW MATERIALS FOR THE HAIR-SPRINGS

The caliber 3255 balance wheel has a blue Parachrom hairspring manufactured by Rolex and made of a patented niobium-zirconium alloy. This nonmagnetic hairspring, presented for the first time in 2000, is temperature stable and is reportedly up to ten times less sensitive to shocks than a conventional hairspring. It is equipped with an optimized Breguet overcoil, ensuring better isochronism of its oscillations in all positions.

This large balance wheel with variable inertia has four Microstella nuts for fine adjustment. Thanks to its reworked geometric shape and more-precise machining, it was possible to improve its sphere of action by a factor of three.

In the future, all the movements for women's watches will be equipped with a hairspring made of synthetic material. "Syloxi" is the Rolex name for this composite material made of silicon and silicon oxide. Its geometry, with its increasing pitch and height, allows for mounting the overcoil at two points to perfectly center the hairspring and keep it flat. This is only clipped onto the balance staff, not glued. The Syloxi hairspring is manufactured in-house using the DRIE (deep reactive ion etching) method and made its debut in 2014, in caliber 2236. It also features a nickel-phosphorus (nonmagnetic) escapement wheel made by the LIGA process and a mainspring developed in-house.

Peter Braun

The Syloxi hairspring for the new generation of women's watch calibers isn't glued on but, rather, clipped onto the balance staff so that it is dimensionally stable. The crescent-shaped "overcoil" is attached at two points and provides stability in the plane of oscillation.

The gleaming, deep-blue Parachrom hairspring is made at Rolex from a niobium-zirconium alloy. The classic curved Breguet overcoil enables the concentric oscillating movement of the hairspring, also called "breathing."

UNITS PER HOUR
ROLEX
COSMOGRAPH
DAYTONA

the Most Beautiful Rolex Classics

On the following pages, with the kind support of Dr. Crott Auctioneers of Mannheim, Germany, we have compiled a selection of the most sought-after Rolex models of the twentieth century and are presenting them with updated estimated values.

1924 Oyster "For all climates"

Case: 9-karat yellow gold, screw-down case back, leather strap, 32 × 34 mm
Movement: Rhodium plated, six adj., manual winding
Remarks: This rare Oyster already drew attention to the fact that it was able to withstand all climatic stresses by its "For all climates" designation.
Estimated value: $1,675

1935 Oyster "Channel Swimmer" Watch

Reference: 3224
Case: Silver, screw-down case back, leather strap, 33 mm
Movement: Rhodium plated, manual winding
Remarks: Called the "Channel Swimmer" in collector circles, this watch recalls when Mercedes Gleitze wore a Rolex Oyster on her wrist when she swam the English Channel in 1927. This was the most impressive endurance test for a waterproof wristwatch up to that date.
Estimated value: $1,117

1932 Oyster Perpetual Chronometer manufactured for Ronchi of Milan

Reference: 3347
Case: 18-karat red gold, screw-down case back, leather strap, 29 mm
Movement: Rhodium plated, rotor winding mechanism
Remarks: Extremely rare, early Oyster with double hour markers; on the bezel, the "III" and "IX" hour markers are in a vertical position, while they are presented lying horizontal on the dial. The watch bears the rare double signature of the jeweler Ronchi of Milan.
Estimated value: $6,700

1940 Oyster Viceroy Chronometer

Reference: 3359
Case: Stainless steel / rose gold, 29 × 38 mm, three-part, screw-down back, stainless-steel / rose gold link bracelet
Movement: Nickel plated, polished, polished screws, "patented Super Balance," 17 jewels, seven adj., manual winding
Remarks: Extremely rare Oyster Viceroy in a two-tone case. The movement has the patented "Super Balance" balance wheel.
Estimated value: $2,792

1941 Oyster "Raleigh"

Reference: 3478
Case: Stainless steel, screw-down case back, leather strap, 29 × 37 mm
Movement: Rhodium-plated, manual winding
Remarks: Rare Oyster Raleigh
Estimated value: $1,340

1943 "Radiomir" Panerai

Reference: 3646
Case: Stainless steel, screw-down case back, leather strap, 47 mm
Movement: Rhodium plated, with striped ornamentation, chatoned, signed "Rolex 17 Rubis," 17 jewels, manual winding
Remarks: Important combat swimmer watch for the Italian navy. The watch was offered with the original leather strap.
Estimated value: $39,089

1930 Prince Railway Jumping Hours "Observatory Quality"

Reference: 1587HS
Case: 18-karat red/white gold, snap-on back, leather strap, 23 × 42 mm
Movement: Rhodium plated, 6 adj., manual winding
Remarks: Rare men's watch in a stepped rectangular case with jumping digital hour display and small seconds
Estimated value: $13,402

1935 Prince Chronometer

Case: 18-karat yellow gold, snap-on back, leather strap, 20 × 40 mm
Movement: Rhodium plated, 8 adj., manual winding
Remarks: Elegant "Prince Classic"
Estimated value: $4,467

1936 Prince Railway "Observatory Quality"

Reference: 1527M
Case: 18-karat white/yellow gold, snap-on case back, leather strap, 22 × 42 mm
Movement: Rhodium plated, 15 jewels, 6 adj., manual winding
Remarks: Rare Prince Railway in perfect state of preservation with Observatory Quality movement
Estimated value: $11,168

1935 Prince Brancard Chronometer "Observatory Quality"

Case: 18-karat white/yellow gold, snap-on case back, leather cord, 26 × 43 mm
Movement: Caliber 971, rhodium plated, 6 adj., manual winding
Remarks: Rare Prince Brancard chronometer with an Observatory Quality movement
Estimated value: $16,752

1945 Prince Brancard Chronometer Jumping Hours "Extra Prima Observatory Quality"

Reference: 1491
Case: 9-karat white gold, snap-on case back, leather strap, 25 × 43 mm
Movement: Rhodium plated, 6 adj., manual winding
Remarks: Rare chronometer with digital jumping hour display. This watch was offered with its original box and a certificate from the observatory in Biel.
Estimated value: $18,986

1950 Prince Aerodynamic Chronometer

Reference: 3361
Case: 18-karat rose gold, 19 × 46 mm, two part, snap-on case back, leather strap
Movement: Caliber 310, rhodium plated, polished, polished screws, "Ultra Prima" quality, 18 jewels, 6 adj.
Remarks: The Prince "Aerodynamic" was presented in 1939. Decreasing case thickness in profile. The caliber 310 was manufactured from 1932 until 1938 but was also encased in ultra prima quality in the later Prince models.
Estimated value: $7,818

1932 Oyster Perpetual Super Precision

Reference: 3353
Case: Stainless steel / gold, screw-down case back, stainless-steel link bracelet, 29 mm
Movement: Rhodium plated, rotor winding mechanism
Remarks: Extremely rare, early automatic Oyster with small seconds
Estimated value: $5,584

1938 Oyster Perpetual Chronometer Bubble Back "Hooded Lugs"

Case: Stainless steel / gold, screw-down case back, leather strap, 32 mm
Movement: Rhodium plated, rotor winding mechanism
Remarks: Oyster Perpetual with Roman numerals and case with hooded lugs
Estimated value: $4,467

1948 Oyster Perpetual Bubble Back

Reference: 3133
Case: Stainless steel, screw-down case back, gold bezel, stainless-steel/gold-link bracelet, 32 mm
Movement: Rhodium plated, rotor winding mechanism
Remarks: Oyster "Bubble Back" with large, easy-to-read luminous numerals
Estimated value: $6,700

1948 Oyster Perpetual Chronometer

Reference: 3372
Case: 18-karat yellow gold, 32 mm, three part, screw-down case back, 18-karat gold link bracelet
Movement: Rhodium plated, polished screws, rotor winding mechanism
Remarks: Rare Oyster "Bubble Back" chronometer with lavender-colored enamel dial and index bezel. The watch was offered with its original box. This trio of watches (see also the next two items) was made for an Indian customer on special request.
Estimated value: All three as a set, $167,523

1948 Oyster Perpetual Chronometer

Reference: 3131
Case: 18-karat yellow gold, 32 mm, three part, screw-down case back, 18-karat Oyster gold link bracelet
Movement: Rhodium plated, polished screws, rotor winding mechanism
Remarks: Rare automatic "Bubble Back" chronometer with blue enamel dial. The watch was offered with its original box.
Estimated value: All three as a set, $167,523

1949 Oyster Perpetual Chronometer

Reference: 3131
Case: 18-karat yellow gold, 32 mm, three part, screw-down case back, 18-karat Oyster gold link bracelet
Movement: Rhodium plated, polished screws, rotor winding mechanism
Remarks: Rare "Bubble Back" chronometer with green enamel dial. The watch was offered with its original box.
Estimated value: All three watches as a set, $167,523

1949 Oyster Perpetual Chronometer "California Dial"

Reference: 5013
Case: Stainless steel, screw-down case back, red gold bezel, leather strap, 32 × 39 mm
Movement: Rhodium plated, rotor winding mechanism
Remarks: Rare Oyster with original "California" dial (half Roman, half Arabic) and small seconds.
Estimated value: $5,584

1949 Oyster Perpetual Chronometer Bubble Back "Hooded Lugs"

Reference: 3065
Case: Stainless steel / gold, screw-down case back, two-tone link bracelet, 32 × 40 mm
Movement: Rhodium plated, rotor winding mechanism
Remarks: Extremely rare "Bubble Back" with hooded lugs and original "California" dial (half Roman, half Arabic). The watch was offered with its original box.
Estimated value: $11,168

1953 Oyster Perpetual Precision Explorer

Reference: 6150
Case: Stainless steel, screw-down case back, leather strap, 35 mm
Movement: Rhodium plated, rotor winding mechanism
Remarks: Rare early Explorer model
Estimated value: $22,336

1953 Oyster Perpetual Chronometer Explorer

Reference: 6350
Case: Stainless steel, screw-down case back, leather strap, 35 mm
Movement: Rhodium plated, rotor winding mechanism
Remarks: Rare early Explorer model with textured dial
Estimated value: $39,089

1956 Oyster Explorer "Officially Certified Chronometer"

Reference: 6610
Case: Stainless steel, 36 mm, three part, screw-down case back, leather strap
Movement: Caliber 1030, rhodium plated, polished, polished screws, rotor winding mechanism, 25 jewels, 6 adj.
Remarks: Rare Oyster Perpetual Chronometer with white dial
Estimated value: $16,752

1958 Oyster Perpetual Explorer Super Precision

Reference: 5500
Case: Stainless steel, screw-down case back, stainless-steel link bracelet, 34 mm
Movement: Caliber 1530, rhodium plated, rotor winding mechanism
Remarks: Rare Explorer
Estimated value: $8,934

1954 Star Dial

Reference: 6098
Case: 18-karat yellow gold, 35 mm, two part, screw-down case back, leather strap
Movement: Red gold plated, polished screws, rotor winding mechanism, 25 jewels
Remarks: "Big Bubble Back" in an unusual combination of a case with the textured "Honeycomb" dial. This has extremely rare appliquéd golden star indices. The watch was offered in its original box.
Estimated value: $22,336

1952 Oyster Perpetual Datejust Chronometer

Reference: 6105
Case: 18-karat yellow gold, screw-down case back, leather strap, 35 mm
Movement: Rhodium plated, rotor winding mechanism
Remarks: Elegant Oyster Datejust in gold. The watch was offered with the original test certificate from the Biel Observatory.
Estimated value: $6,143

1955 Oyster Perpetual Datejust Superlative Chronometer "Serpico y Laino"

Reference: 6605
Case: 18-karat yellow gold, screw-down case back, leather strap, 36 mm
Movement: Caliber 1065, rhodium plated, 25 jewels, 6 adj., rotor winding mechanism
Remarks: Oyster with extremely rare dial imprint "Superlative Chronometer by Official Test, 50 m = 165 ft" and the second signature of Serpico y Laino
Estimated value: $6,143

1945 Oyster Perpetual Chronometer

Reference: 4467
Case: 18-karat yellow gold, screw-down case back, leather strap, 36 mm
Movement: Rhodium plated, rotor winding mechanism
Remarks: Elegant Oyster in gold case
Estimated value: $11,168

1955 Datejust Chronometer

Reference: 6305/1
Case: Stainless steel, screw-down case back, stainless-steel link bracelet, 36 mm
Movement: Caliber 745, rhodium plated, rotor winding mechanism
Remarks: Rare Datejust with guilloche dial, black-red date display, and white gold bezel. This model was produced only from 1953 until 1955.
Estimated value: $4,467

1956 Date Chronometer

Reference: 1503
Case: 18-karat yellow gold, screw-down case back, golden Milanaise mesh bracelet, 34 mm
Movement: Caliber 1560, rhodium plated, decorated, 26 jewels, 6 adj., rotor winding mechanism
Remarks: Rolex Date with double signature, Rolex and "Serpico y Laino."
Estimated value: $5,584

1958 Oyster Perpetual "Milgauss" Superlative Chronometer "Officially Certified"

Reference: 6541
Case: Stainless steel, screw-down case back, rotating bezel, stainless-steel link bracelet, 36 mm
Movement: Caliber 1066M, rhodium plated, 25 jewels, 6 adj., rotor winding mechanism
Remarks: Extremely rare Milgauss of the first generation, still without a "lightning pointer"
Estimated value: $67,009

1966 Oyster Perpetual "Milgauss"

Reference: 1019
Case: Stainless steel, screw-down case back, stainless-steel link bracelet, 37 mm
Movement: Rhodium plated, automatic self-winding
Remarks: Rare early Milgauss. The movement is surrounded by a soft-iron protective cap and thus protected against magnetism up to 1000 Gauss.
Estimated value: $27,920

1972 Oyster Perpetual Datejust Chronometer "Thunderbird"

Reference: 1625
Case: 18-karat yellow gold, screw-down case back, leather strap, 36 mm
Movement: Caliber 1570, rhodium plated, rotor winding mechanism
Remarks: Early Thunderbird in gold
Estimated value: $6,700

1977 Oyster Perpetual Datejust "Thunderbird" Chronometer

Reference: 1625
Case: Stainless steel, screw-down case back, 18-karat gold bezel, leather strap, 36 mm
Movement: Rhodium plated, rotor winding mechanism
Remarks: Rare Oyster Thunderbird with black dial
Estimated value: $4,467

1954 Oyster Perpetual Turn-O-Graph

Reference: 6202
Case: Stainless steel, 36 mm, three part, rotating bezel, screw-down case back, leather strap
Movement: nickel plated, polished, polished screws, rotor winding mechanism
Remarks: Early Oyster Turn-O-Graph with rotating black bezel with indices
Estimated value: $22,336

1956 Oyster Perpetual GMT-Master Chronometer "Officially Certified"

Reference: 6542
Case: Stainless steel, screw-down case back, rotating bezel, stainless-steel link bracelet, 38 mm
Movement: Caliber 1030, rhodium plated, 25 jewels, 6 adj., rotor winding mechanism
Remarks: This watch has its original acrylic rotating bezel. A second time zone can be set using the 24-hour hand and the bezel.
Estimated value: $89,346

1955 Oyster Perpetual Submariner "James Bond"

Reference: 6205
Case: Stainless steel, screw-down case back, stainless-steel link bracelet, 36 mm
Movement: Caliber A260, rhodium plated, rotor winding mechanism
Remarks: This watch got its nickname from its appearances in several early James Bond films.
Estimated value: $50,257

1957 Oyster Perpetual Submariner 100 m / 330 ft

Reference: 6536/1
Case: Stainless steel, screw-down case back, rotating bezel, stainless-steel link bracelet, 37 mm
Movement: Caliber 1030, rhodium plated, 25 jewels, 6 adj., rotor winding mechanism
Remarks: Early Chronometer model of the Submariner
Estimated value: $22,336

1958 Oyster Perpetual Submariner 200 m / 660 ft "James Bond"

Reference: 6538
Case: Stainless steel, screw-down case back, stainless-steel link bracelet, 37 mm
Movement: Caliber 1030, rhodium plated, 25 jewels, rotor winding mechanism
Remarks: This Submariner owes its nickname to its appearances at several early James Bond films.
Estimated value: $111,682

1959 Oyster Perpetual Submariner "Officially Certified Chronometer"

Reference: 6538
Case: Stainless steel, screw-down case back, rotating bezel, stainless-steel link bracelet, 38 mm
Movement: Caliber 1030, rhodium plated, 25 jewels, 6 adj., rotor winding mechanism
Remarks: Unique Submariner with its original dial in an almost unknown version, with red depth indicator and "Officially Certified Chronometer" in red lettering. This is also a so-called James Bond model.
Estimated value: $167,523

1964 Oyster Perpetual Submariner "200 m = 660 ft" Explorer dial

Reference: 5513
Case: Stainless steel, screw-down case back, stainless-steel link bracelet, 39 mm
Movement: Caliber 1520, rhodium plated, 26 jewels, 6 adj., rotor winding mechanism
Remarks: Submariner with an extremely rare dial design in the "Explorer" style
Estimated value: $67,009

1970 Oyster Perpetual Submariner

Reference: 5513
Case: Stainless steel, 39 mm, three part, rotating bezel, screw-down case back, textile diving strap
Movement: Rhodium plated, polished, polished screws, rotor winding mechanism, 26 jewels
Remarks: Military combat swimmer watch of the "British Special Boat Service" of the Royal Marines. The original sword hands shown here are a distinguishing feature of this extremely rare watch. In addition, the watches had to have fixed bar lugs.
Estimated value: $111,682

1950 Perpetual Chronometer Precision

Reference: 8171
Case: Stainless steel, screw-down case back, leather strap, 38 mm
Movement: Rhodium plated, rotor winding mechanism
Remarks: Extremely rare Oyster with full calendar, small seconds, and moon phase. This model was manufactured only in a small series and is very difficult to find. This watch bears the number 279.
Estimated value: $279,205

1949 Perpetual Chronometer

Reference: 8171
Case: 18-karat red gold, snap-on case back, leather strap, 38 mm
Movement: Rhodium plated, rotor winding mechanism
Remarks: Extremely rare men's watch that was manufactured in a small series. This watch (serial number 202) is almost impossible to find in rose gold. This watch was offered with its original box.
Estimated value: $111,682

1953 Oyster Perpetual Chronometer "Officially Certified"

Reference: 6062
Case: Stainless steel, screw-down case back, stainless-steel link bracelet, 36 mm
Movement: Rhodium plated, rotor winding mechanism
Remarks: Extremely rare automatic Oyster with full calendar and moon phase. Only 350 pieces were manufactured between 1950 and 1953 in yellow gold, 50 pieces in red gold, and an extremely small number in stainless steel.
Estimated value: $167,523

1949 Dato Compax Chronograph

Reference: 4768
Case: Stainless steel, 35 mm, three part, snap-on case back, leather strap
Movement: Caliber 72C, rhodium plated, polished, with column wheel control, finely polished, beveled chronograph steel parts, mirror-polished screws, 17 jewels, manual winding
Remarks: Rare chronograph with 30-minute and 12-hour counter. Full calendar with pointer date display, day of the week, and month in the window. Only 220 pieces were manufactured.
Estimated value: $67,009

1953 Oyster Chronograph

Reference: 6236
Case: 18-karat yellow gold, 36 mm, three part, screw-down case back, leather strap
Movement: Caliber 72C, rhodium plated, polished, with column wheel control, finely polished, beveled chronograph steel parts, mirror-polished screws, manual winding
Remarks: Extremely rare chronograph with full calendar. In collectors' circles, the Reference 6236 is also known as the "Jean-Claude Killy." Only 170 pieces made in yellow gold.
Estimated value: $134,018

1954 Oyster Chronograph Antimagnetic "Jean-Claude Killy"

Reference: 6036
Case: Stainless steel, screw-down case back, stainless-steel link bracelet, 36 mm
Movement: Caliber 72C, rhodium plated, with column wheel control, manual winding
Remarks: Rare chronograph with 30-minute and 12-hour counters as well as full calendar with date, day of the week, and month. Model named after the Alpine ski racer of the same name.
Estimated value: $223,364

1932 Military Antimagnetic Chronograph

Reference: 2508
Case: Stainless steel, 35 mm, three part, snap-on case back, leather strap
Movement: Nickel plated, polished, with column wheel control, finely polished chronograph steel parts, polished screws, manual winding
Remarks: Rare chronograph with a 30-minute counter. This chronograph is one of the first models produced by Rolex with a timer.
Estimated value: $44,673

1935 Chronograph

Reference: 3371
Case: 18-karat yellow gold, 35 mm, three part, snap-on case back, leather strap
Movement: nickel plated, polished, with column wheel control, finely polished chronograph steel parts, polished screws, 17 jewels, manual winding
Remarks: Rare, extremely fine chronograph with a 30-minute counter as well as tachymetric and blue telemeter scales. This watch was offered with a Rolex gold clasp.
Estimated value: $16,752

1934 Chronograph Antimagnetic

Reference: 2508
Case: 18-karat yellow gold, snap-on case back, leather strap, 36 mm
Movement: Rhodium plated, with column wheel control, manual winding
Remarks: One of the first chronographs produced by Rolex with 30-minute counter and tachymetric scale
Estimated value: $16,752

1934 Chronograph Anti-Magnétique

Reference: 3834
Case: 18-karat yellow gold, snap-on case back, leather strap, 32 mm
Movement: Rhodium plated, manual winding
Remarks: Extremely rare chronograph with 30-minute counter and additional tachymetric scale
Estimated value: $13,402

1947 Oyster Chronograph

Reference: 4500
Case: Stainless steel, screw-down case back, leather strap, 36 mm
Movement: Caliber R23, rhodium plated, with column wheel control, manual winding
Remarks: Rare Oyster chronograph with 30-minute counter and additional tachymetric scale
Estimated value: $39,089

1945 Chronograph Antimagnetic

Reference: 3525
Case: 18-karat yellow gold, screw-down case back, leather strap, 35 mm
Movement: Rhodium plated, with manual winding, column wheel control
Remarks: One of the first Oyster chronographs manufactured by Rolex in 1945 with a 30-minute counter and telemetric scale. The watch was offered with its original box.
Estimated value: $55,841

1965 Chronograph ("Pre-Daytona")

Case: 14-karat yellow gold, screw-down case back, leather strap, 36 mm
Movement: Rhodium plated, with column wheel control, manual winding
Remarks: Blue tachymetric scale
Estimated value: $50,257

1966 Chronograph ("Pre-Daytona")

Reference: 6238
Case: Stainless steel, screw-down case back, stainless-steel link bracelet, 36 mm
Movement: Caliber 722.1, rhodium plated, with column wheel control, 17 jewels, 3 adj., manual winding
Remarks: Chronograph with 30-minute and 12-hour counters
Estimated value: $50,257

1965 Chronograph ("Pre-Daytona")

Reference: 6238
Case: 14-karat yellow gold, screw-down case back, yellow gold link bracelet, 32 mm
Movement: Caliber 72B, rhodium plated, with column wheel control, manual winding
Remarks: Chronograph with 30-minute and 12-hour counters
Estimated value: $111,682

1960 Oyster Cosmograph Daytona "Paul Newman"

Reference: 6240
Case: Stainless steel, screw-down case back, stainless-steel link bracelet, 37 mm
Movement: Rhodium plated, with column wheel control, manual winding
Remarks: Rare "Daytona" chronograph with black tachymetric scale on the bezel. The watch was offered with proof of overhaul and confirmation of originality.
Estimated value: $223,364

1967 Oyster Cosmograph Daytona "Paul Newman"

Reference: 6239
Case: Stainless steel, screw-down case back, stainless-steel link bracelet, 36 mm
Movement: With column wheel control, manual winding
Remarks: "Daytona" chronograph with "Paul Newman" dial. The watch was offered with its original box and certificate.
Estimated value: $167,523

1968 Oyster Cosmograph Daytona

Reference: 6239/6263
Case: Stainless steel, screw-down case back, stainless-steel link bracelet, 37 mm
Movement: Caliber 727, rhodium plated, with column wheel control, manual winding
Remarks: Rare "Cosmograph" with 30-minute and 12-hour counters and additional tachymetric scale on the bezel
Estimated value: $78,177

the Models of the Current Collection

All prices given are as of September 1, 2020.

The following overview primarily shows a selection of the current men's watch collection from Rolex, with a focus on the "tool watches."

Oyster Perpetual GMT-Master II

Reference: 126710BLNR
Movement: Self-winding automatic, Rolex caliber 3285; 28.5 mm diameter, 6.4 mm height; 31 jewels; 28,800 vph; Parachrom hairspring, Paraflex shock absorber, Chronergy escapement; power reserve 70 hours; certified chronometer (COSC)
Functions: Hours (adjustable in increments via the crown), minutes, center seconds; additional 24-hour display (second time zone); date
Case: Stainless steel; 40 mm diameter, 13 mm thickness; bezel with ceramic insert, bidirectionally rotatable, 24 index scale; sapphire crystal; screw-down crown; waterproof to 10 bar
Bracelet: Jubilee stainless steel, folding clasp, with safety catch and extension link
Price: $9,828

Oyster Perpetual GMT-Master II

Reference: 126710BLRO
Movement: Self-winding automatic, Rolex caliber 3285; 28.5 mm diameter, 6.4 mm height; 31 jewels; 28,800 vph; Parachrom hairspring, Paraflex shock absorber, Chronergy escapement; power reserve 70 hours; certified chronometer (COSC)
Functions: Hours (adjustable in increments via the crown), minutes, center seconds; additional 24-hour display (second time zone); date
Case: Stainless steel; 40 mm diameter; 13 mm thickness; bezel with ceramic insert, bidirectionally rotatable, 24 index scale; sapphire crystal; screw-down crown; waterproof to 10 bar
Bracelet: Jubilee stainless steel, folding clasp, with safety catch and extension link
Price: $9,828

Oyster Perpetual GMT-Master II

Reference: 126719BLRO
Movement: Self-winding automatic, Rolex caliber 3285; 28.5 mm diameter, 6.4 mm height; 31 jewels; 28,800 vph; Parachrom hairspring, Paraflex shock absorber, Chronergy escapement; power reserve 70 hours; certified chronometer (COSC)
Functions: Hours (adjustable in increments via the crown), minutes, center seconds; additional 24-hour display (second time zone); date
Case: White gold; 40 mm diameter, 13 mm thickness; bezel with ceramic insert, bidirectionally rotating, with 24 index scale; sapphire crystal; screw-down crown; waterproof to 10 bar
Bracelet: Oyster white gold, folding clasp, with safety catch and extension link
Price: $39,089

Oyster Perpetual GMT-Master II

Reference: 126711CHNR
Movement: Self-winding automatic, Rolex caliber 3285; 28.5 mm diameter, 6.4 mm height; 31 jewels; 28,800 vph; Parachrom hairspring, Chronergy escapement; power reserve 70 hours; certified chronometer (COSC)
Functions: Hours (adjustable in increments via the crown), minutes, center seconds; additional 24-hour display (second time zone); date
Case: Stainless steel; 40 mm diameter, 13 mm thickness; bezel in rose gold with ceramic insert, bidirectionally rotatable, with 24 index scale; sapphire crystal; screw-down crown; waterproof to 10 bar
Bracelet: Oyster stainless steel with rose gold elements, folding clasp, with safety catch and extension link
Price: $15,077

Oyster Perpetual GMT-Master II

Reference: 126715CHNR
Movement: Self-winding automatic, Rolex caliber 3285; 28.5 mm diameter, 6.4 mm height; 31 jewels; 28,800 vph; Parachrom hairspring, Chronergy escapement; power reserve 70 hours; certified chronometer (COSC)
Functions: Hours (adjustable in increments via the crown), minutes, center seconds; additional 24-hour display (second time zone); date
Case: Rose gold; 40 mm diameter, 13 mm thickness; bezel in rose gold with ceramic insert, bidirectionally rotatable, with 24 index scale; sapphire crystal; screw-down crown; waterproof to 10 bar
Bracelet: Oyster rose gold, folding clasp, with safety catch and extension link
Price: $39,089

Oyster Perpetual GMT-Master II

Reference: 126719BLRO
Movement: Self-winding automatic, Rolex caliber 3285; 28.5 mm diameter, 6.4 mm height; 31 jewels; 28,800 vph; Parachrom hairspring, Chronergy escapement; power reserve 70 hours; certified chronometer (COSC)
Functions: Hours (adjustable in increments via the crown), minutes, center seconds; additional 24-hour display (second time zone); date
Case: White gold; 40 mm diameter, 13 mm thickness; bezel with ceramic insert, bidirectionally rotating, with 24 index scale; sapphire crystal; screw-down crown; waterproof to 10 bar
Bracelet: Oyster white gold, folding clasp, with safety catch and extension link
Remarks: Meteorite dial.
Price: $40,764

Oyster Perpetual Deepsea

Reference: 126660
Movement: Self-winding, Rolex caliber 3235; 29.1 mm diameter; 31 jewels; 28,800 vph; Parachrom hairspring, Chronergy escapement; power reserve 70 hours; certified chronometer (COSC)
Functions: Hours, minutes, center seconds; date
Case: Stainless steel; 44 mm diameter; bezel with ceramic insert, unidirectionally rotatable, with 60-minute scale; sapphire crystal; screw-down crown; helium valve; waterproof to 390 bar
Bracelet: Oyster stainless steel, folding clasp, with safety latch, extension link and fine adjustment
Price: $13,123

Oyster Perpetual Deepsea

Reference: 126660
Movement: Self-winding, Rolex caliber 3235; 29.1 mm diameter; 31 jewels; 28,800 vph; Parachrom hairspring, Paraflex shock absorber, Chronergy escapement, Glucydur balance wheel with Microstella regulating nuts; power reserve 70 hours; certified chronometer (COSC)
Functions: Hours, minutes, center seconds; date
Case: Stainless steel; 44 mm diameter; bezel with ceramic insert, unidirectionally rotatable, with 60-minute scale; sapphire crystal; screw-down crown; helium valve; waterproof to 390 bar
Bracelet: Oyster stainless steel, folding clasp, with safety latch, extension link and fine adjustment
Price: $12,843

Oyster Perpetual Sea-Dweller

Reference: 126603
Movement: Self-winding, Rolex caliber 3235; 29.1 mm diameter; 31 jewels; 28,800 vph; Parachrom hairspring, Paraflex shock absorber, Chronergy escapement, Glucydur balance wheel with Microstella regulating nuts; power reserve 70 hours; certified chronometer (COSC)
Functions: Hours, minutes, center seconds; date
Case: Stainless steel; 43 mm diameter, 13.8 mm thickness; bezel in yellow gold with ceramic insert, unidirectionally rotatable, with 60-minute scale; sapphire crystal; screw-down crown; helium valve; water resistant to 122 bar
Bracelet: Oyster stainless steel with yellow gold elements, folding clasp, with safety catch, extension and fine adjustment
Price: $16,922
Other versions: In stainless steel ($11,894)

Oyster Perpetual Submariner

Reference: 124060
Movement: Self-winding, Rolex caliber 3230; 29.1 mm diameter; 31 jewels; 28,800 vph; Parachrom hairspring, Paraflex shock absorber, Chronergy escapement, Glucydur balance wheel with Microstella regulating nuts; power reserve 70 hours; certified chronometer (COSC)
Functions: Hours, minutes, center seconds
Case: Stainless steel; 41 mm diameter, 12.5 mm thickness; bezel with ceramic insert, unidirectionally rotatable, with 60-minute scale; sapphire crystal; screw-down crown; waterproof to 30 bar
Bracelet: Oyster stainless steel, folding clasp, with extension link
Price: $8,209

Oyster Perpetual Submariner Date

Reference: 126610LV
Movement: Self-winding, Rolex caliber 3235; 29.1 mm diameter; 31 jewels; 28,800 vph; Parachrom hairspring, Paraflex shock absorber, Chronergy escapement, Glucydur balance wheel with Microstella regulating nuts; power reserve 70 hours; certified chronometer (COSC)
Functions: Hours, minutes, center seconds; date
Case: Stainless steel; 41 mm diameter, 12.5 mm thickness; bezel with ceramic insert, unidirectionally rotatable, with 60-minute scale; sapphire crystal; screw-down crown; waterproof to 30 bar
Bracelet: Oyster stainless steel, folding clasp, with extension link
Price: $9,716
Other versions: With black Cerachrom bezel ($9,270)

Oyster Perpetual Submariner Date

Reference: 126619LB
Movement: Self-winding, Rolex caliber 3235; 29.1 mm diameter; 31 jewels; 28,800 vph; Parachrom hairspring, Paraflex shock absorber, Chronergy escapement, Glucydur balance wheel with Microstella regulating nuts; power reserve 70 hours; certified chronometer (COSC)
Functions: Hours, minutes, center seconds; date
Case: White gold; 40 mm diameter, 12.5 mm thickness; bezel with ceramic insert, unidirectionally rotatable, with 60-minute scale; sapphire crystal; screw-down crown; waterproof to 30 bar
Bracelet: Oyster white gold, folding clasp, with extension link
Price: $40,429

Oyster Perpetual Air-King

Reference: 116900
Movement: Self-winding, Rolex caliber 3131; 28.5 mm diameter; 31 jewels; 28,800 vph; Parachrom hairspring, Glucydur balance wheel with Microstella regulating nuts; magnetic-field resistance via soft-iron inner cage; power reserve 48 hours; certified chronometer (COSC)
Functions: Hours, minutes, center seconds
Case: Stainless steel; 40 mm diameter; sapphire crystal; screw-down crown; waterproof to 10 bar
Bracelet: Oyster stainless steel, folding clasp, with safety catch and extension link
Price: $6,533

Oyster Perpetual Milgauss

Reference: 116400GV
Movement: Self-winding, Rolex caliber 3131; 28.5 mm diameter; 31 jewels; 28,800 vph; Parachrom hairspring, Glucydur balance wheel with Microstella regulating nuts; magnetic-field resistance via soft-iron inner cage; power reserve 48 hours; certified chronometer (COSC)
Functions: Hours, minutes, center seconds
Case: Stainless steel; 40 mm diameter; sapphire crystal; screw-down crown; waterproof to 10 bar
Bracelet: Oyster stainless steel, folding clasp, with safety catch and extension link
Price: $8,488

Oyster Perpetual Milgauss

Reference: 116400GV
Movement: Self-winding, Rolex caliber 3131; 28.5 mm diameter; 31 jewels; 28,800 vph; Parachrom hairspring, Glucydur balance wheel with Microstella regulating nuts; magnetic-field resistance via soft-iron inner cage; power reserve 48 hours; certified chronometer (COSC)
Functions: Hours, minutes, center seconds
Case: Stainless steel; 40 mm diameter; sapphire crystal; screw-down crown; waterproof to 10 bar
Bracelet: Oyster stainless steel, folding clasp, with safety catch and extension link
Price: $8,488

Oyster Perpetual Explorer

Reference: 214270
Movement: Self-winding, Rolex caliber 3132; 28.5 mm diameter; 31 jewels; 28,800 vph; Parachrom hairspring, Paraflex shock absorber, Glucydur balance wheel with Microstella regulating nuts; power reserve 48 hours; certified chronometer (COSC)
Functions: Hours, minutes, center seconds
Case: Stainless steel; 39 mm diameter; sapphire crystal; screw-down crown; waterproof to 10 bar
Bracelet: Oyster stainless steel, folding clasp, with extension link
Price: $6,645

Oyster Perpetual Explorer II

Reference: 216570
Movement: Self-winding, Rolex caliber 3187; 28.5 mm diameter; 31 jewels; 28,800 vph; Parachrom Breguet hairspring; power reserve 48 hours; certified chronometer (COSC)
Functions: Hours (can be adjusted in increments via the crown), minutes, center seconds; additional 24 hour display (second time zone); date
Case: Stainless steel; 42 mm diameter; sapphire crystal; screw-down crown; waterproof to 10 bar
Bracelet: Oyster stainless steel, folding clasp, with extension link
Price: $8,488

Oyster Perpetual Explorer II

Reference: 216570
Movement: Self-winding, Rolex caliber 3187; 28.5 mm diameter; 31 jewels; 28,800 vph; Parachrom Breguet hairspring; power reserve 48 hours; certified chronometer (COSC)
Functions: Hours (can be adjusted in increments via the crown), minutes, center seconds; additional 24 hour display (second time zone); date
Case: Stainless steel; 42 mm diameter; sapphire crystal; screw-down crown; waterproof to 10 bar
Bracelet: Oyster stainless steel, folding clasp, with extension link
Price: $8,488

Oyster Perpetual Cosmograph Daytona

Reference: 116500LN
Movement: Self-winding, Rolex caliber 4130; 30.5 mm diameter, 6.5 mm height; 44 jewels; 28,800 vph; Parachrom hairspring, Glucydur balance wheel with Microstella regulating nuts; power reserve 72 hours; certified chronometer (COSC)
Functions: Hours, minutes, small seconds; chronograph
Case: Stainless steel; 40 mm diameter; thickness 12.8 mm; Cerachrom bezel; sapphire crystal; screw-down crown and pushers; waterproof to 10 bar
Bracelet: Oyster stainless steel, folding clasp, with safety catch and extension link
Price: $13,325
Other versions: With black dial

Oyster Perpetual Cosmograph Daytona

Reference: 116500LN
Movement: Self-winding, Rolex caliber 4130; 30.5 mm diameter, 6.5 mm height; 44 jewels; 28,800 vph; Parachrom hairspring, Glucydur balance wheel with Microstella regulating nuts; power reserve 72 hours; certified chronometer (COSC)
Functions: Hours, minutes, small seconds; chronograph
Case: Stainless steel; 40 mm diameter, 12.8 mm thickness; Cerachrom bezel; sapphire crystal; screw-down crown and pushers; waterproof to 10 bar
Bracelet: Oyster stainless steel, folding clasp, with safety catch and extension link
Price: $13,325
Other versions: With white dial

Oyster Perpetual Cosmograph Daytona

Reference: 116503
Movement: Self-winding, Rolex caliber 4130; 30.5 mm diameter, 6.5 mm height; 44 jewels; 28,800 vph; Parachrom hairspring, Glucydur balance wheel with Microstella regulating nuts; power reserve 72 hours; certified chronometer (COSC)
Functions: Hours, minutes, small seconds; chronograph
Case: Stainless steel; 40 mm diameter, 12.8 mm thickness; bezel in yellow gold; sapphire crystal; screw-down crown and pushers; waterproof to 10 bar
Bracelet: Oyster stainless steel with yellow gold elements, folding clasp, with safety catch and extension link
Price: $17,702

Oyster Perpetual Sky-Dweller

Reference: 326934
Movement: Self-winding, Rolex caliber 9001; 33 mm diameter, 8 mm height; 40 jewels; 28,800 vph; power reserve 72 hours; certified chronometer (COSC)
Functions: Hours, minutes, center seconds; additional 24-hour display (second time zone); annual calendar with date, month
Case: Stainless steel; 42 mm diameter, 14.1 mm thickness; bezel in white gold, bidirectionally rotatable to control the functions; sapphire crystal; screw-down crown; waterproof to 10 bar
Bracelet: Oyster stainless steel, folding clasp, with extension link
Price: $15,133
Other versions: Yellow gold ($47,388); rose gold (on request)

Oyster Perpetual Sky-Dweller

Reference: 326933
Movement: Self-winding, Rolex caliber 9001; 33 mm diameter, 8 mm height; 40 jewels; 28,800 vph; Parachrom hairspring, Paraflex shock absorber, Glucydur balance wheel with Microstella regulating nuts; power reserve 72 hours; certified chronometer (COSC)
Functions: Hours, minutes, center seconds; additional 24-hour display (second time zone); annual calendar with date, month
Case: Stainless steel; 42 mm diameter, 14.1 mm thickness; yellow gold bezel, bidirectional rotatable to control the functions; sapphire crystal; screw-down crown; waterproof to 10 bar
Bracelet: Oyster stainless steel with yellow gold elements, folding clasp, with extension link
Price: $17,981

Oyster Perpetual Sky-Dweller

Reference: 326238
Movement: Self-winding, Rolex caliber 9001; 33 mm diameter, 8 mm height; 40 jewels; 28,800 vph; Parachrom hairspring, Paraflex shock absorber, Glucydur balance wheel with Microstella regulating nuts; power reserve 72 hours; certified chronometer (COSC)
Functions: Hours, minutes, center seconds; additional 24-hour display (second time zone); annual calendar with date, month
Case: Yellow gold; 42 mm diameter, 14.1 mm thickness; bidirectionally rotatable bezel to control the functions; sapphire crystal; screw-down crown; waterproof to 10 bar
Bracelet: Oysterflex (elastomer with flexible metal core), folding clasp, with extension link
Price: $40,652

Oyster Perpetual Yacht-Master II

Reference: 116680
Movement: Self-winding, Rolex caliber 4161 (basic caliber 4130); 31.2 mm diameter, 8.05 mm height; 42 jewels; 28,800 vph; Parachrom hairspring; power reserve 72 hours; certified chronometer (COSC)
Functions: Hours, minutes, small seconds; programmable regatta countdown with memory
Case: Stainless steel; 44 mm diameter, 13.8 mm thickness; bezel with ceramic insert, bidirectionally rotatable to control the functions; sapphire crystal; screw-down crown; waterproof to 10 bar
Bracelet: Oyster stainless steel, folding clasp, with safety catch and extension link
Price: $18,986
Other versions: in yellow gold ($44,338)

Oyster Perpetual Yacht-Master II

Reference: 116681
Movement: Self-winding, Rolex caliber 4161 (base caliber 4130); 31.2 mm diameter, 8.05 mm height; 42 jewels; 28,800 vph; Parachrom hairspring, Glucydur balance wheel with Microstella regulating nuts; power reserve 72 hours; certified chronometer (COSC)
Functions: Hours, minutes, small seconds; programmable regatta countdown with memory
Case: Stainless steel; 44 mm diameter, 13.8 mm thickness; bezel in rose gold with ceramic insert, bidirectionally rotatable; sapphire crystal; screw-down crown; waterproof to 10 bar
Bracelet: Oyster stainless steel with rose gold elements, folding clasp, with safety catch and extension link
Price: $25,799
Other versions: in white gold and platinum ($49,028)

Oyster Perpetual Yacht-Master 42

Reference: 226659
Movement: Self-winding, Rolex caliber 3235; 29.1 mm diameter; 31 jewels; 28,800 vph; Parachrom hairspring, Paraflex shock absorber, Chronergy escapement, Glucydur balance wheel with Microstella regulating nuts; power reserve 70 hours; certified chronometer (COSC)
Functions: Hours, minutes, center seconds; date
Case: White gold; 42 mm diameter, 11.8 mm thickness; bezel with ceramic insert, bidirectionally rotatable, with 60-minute scale; sapphire crystal; screw-down crown; waterproof to 10 bar
Bracelet: Oysterflex (elastomer with flexible metal core), folding clasp, with extension link
Price: $29,372

Oyster Perpetual Yacht-Master 40

Reference: 126655
Movement: Self-winding, Rolex caliber 3235; 29.1 mm diameter; 31 jewels; 28,800 vph; Parachrom hairspring, Paraflex shock absorber, Chronergy escapement, Glucydur balance wheel with Microstella regulating nuts; power reserve 70 hours; certified chronometer (COSC)
Functions: Hours, minutes, center seconds; date
Case: Rose gold; 40 mm diameter, 11.7 mm thickness; bezel with ceramic insert, bidirectionally rotatable, with 60-minute scale; sapphire crystal; screw-down crown; waterproof to 10 bar
Bracelet: Oysterflex (elastomer with flexible metal core), folding clasp, with extension link
Price: $27,809

Oyster Perpetual Yacht-Master 40

Reference: 116622
Movement: Self-winding, Rolex caliber 3235; 29.1 mm diameter; 31 jewels; 28,800 vph; Parachrom hairspring, Paraflex shock absorber, Chronergy escapement, Glucydur balance wheel with Microstella regulating nuts; power reserve 70 hours; certified chronometer (COSC)
Functions: Hours, minutes, center seconds; date
Case: Stainless steel; 40 mm diameter, 11.7 mm thickness; bezel in platinum, rotatable in both directions, with 60-minute scale; sapphire crystal; screw-down crown; waterproof to 10 bar
Bracelet: Oyster stainless steel, folding clasp, with extension link
Price: $12,173

Oyster Perpetual Yacht-Master 37

Reference: 268655
Movement: Self-winding, Rolex caliber 2236; 20 mm diameter, 5.5 mm height; 31 jewels; 28,800 vph; Parachrom hairspring, Glucydur balance wheel with Microstella regulating nuts; power reserve 48 hours; certified chronometer (COSC)
Functions: Hours, minutes, center seconds; date
Case: Rose gold; 37 mm diameter, 11 mm thickness; bezel with ceramic insert, bidirectionally rotatable, with 60-minute scale; sapphire crystal; screw-down crown; waterproof to 10 bar
Bracelet: Oysterflex (elastomer with flexible metal core), folding clasp, with extension link
Price: $23,621

Oyster Perpetual Day-Date 40

Reference: 228239
Movement: Self-winding, Rolex caliber 3255; 29.1 mm diameter, 5.4 mm height; 31 jewels; 28,800 vph; Parachrom hairspring, Paraflex shock absorber, Chronergy escapement, Glucydur balance wheel with Microstella regulating nuts; power reserve 70 hours; certified chronometer (COSC)
Functions: Hours, minutes, center seconds; date and day of the week
Case: White gold; 40 mm diameter, 11.6 mm thickness; sapphire crystal; screw-down crown; waterproof to 10 bar
Bracelet: President white gold, hidden folding clasp
Price: $39,870
Other versions: in rose gold ($39,870)

Oyster Perpetual Day-Date 40

Reference: 228238
Movement: Self-winding, Rolex caliber 3255; 29.1 mm diameter, 5.4 mm height; 31 jewels; 28,800 vph; Parachrom hairspring, Paraflex shock absorber, Chronergy escapement, Glucydur balance wheel with Microstella regulating nuts; power reserve 70 hours; certified chronometer (COSC)
Functions: Hours, minutes, center seconds; date and day of the week
Case: Yellow gold; 40 mm diameter, 11.6 mm thickness; sapphire crystal; screw-down crown; waterproof to 10 bar
Bracelet: President in yellow gold, hidden folding clasp
Price: $37,190

Oyster Perpetual Day-Date 40

Reference: 228206
Movement: Self-winding, Rolex caliber 3255; 29.1 mm diameter, 5.4 mm height; 31 jewels; 28,800 vph; Parachrom hairspring, Paraflex shock absorber, Chronergy escapement, Glucydur balance wheel with Microstella regulating nuts; power reserve 70 hours; certified chronometer (COSC)
Functions: Hours, minutes, center seconds; date and day of the week
Case: Platinum; 40 mm diameter, 11.6 mm thickness; sapphire crystal; screw-down crown; waterproof to 10 bar
Bracelet: President in platinum, hidden folding clasp
Price: On request

Oyster Perpetual Day-Date 36

Reference: 128238
Movement: Self-winding, Rolex caliber 3255; 29.1 mm diameter, 5.4 mm height; 31 jewels; 28,800 vph; Parachrom hairspring, Paraflex shock absorber, Chronergy escapement, Glucydur balance wheel with Microstella regulating nuts; power reserve 70 hours; certified chronometer (COSC)
Functions: Hours, minutes, center seconds; date and day of the week
Case: Yellow gold; 36 mm diameter, 11.1 mm thickness; sapphire crystal; screw-down crown; waterproof to 10 bar
Bracelet: President in yellow gold, folding clasp
Remarks: Dial with diamond hour markers
Price: $36,799
Other versions: In rose gold ($39,591)

Oyster Perpetual Day-Date 36

Reference: 128239
Movement: Self-winding, Rolex caliber 3255; 29.1 mm diameter, 5.4 mm height; 31 jewels; 28,800 vph; Parachrom hairspring, Paraflex shock absorber, Chronergy escapement, Glucydur balance wheel with Microstella regulating nuts; power reserve 70 hours; certified chronometer (COSC)
Functions: Hours, minutes, center seconds; date and day of the week
Case: White gold; 36 mm diameter, 11.1 mm thickness; sapphire crystal; screw-down crown; waterproof to 10 bar
Bracelet: President in white gold, folding clasp
Remarks: Dial with diamond hour markers
Price: $39,591

Oyster Perpetual Day-Date 36

Reference: 128349RBR
Movement: Self-winding, Rolex caliber 3255; 29.1 mm diameter, 5.4 mm height; 31 jewels; 28,800 vph; Parachrom hairspring, Paraflex shock absorber, Chronergy escapement, Glucydur balance wheel with Microstella regulating nuts; power reserve 70 hours; certified chronometer (COSC)
Functions: Hours, minutes, center seconds; date and day of the week
Case: White gold; 36 mm diameter, 11.1 mm thickness; sapphire crystal; screw-down crown; waterproof to 10 bar
Bracelet: President in white gold, folding clasp
Remarks: Mother-of-pearl dial with diamond hour markers
Price: On request

Oyster Perpetual 41

Reference: 124300
Movement: Self-winding, Rolex caliber 3230; 29.1 mm diameter; 31 jewels; 28,800 vph; Parachrom hairspring, Paraflex shock absorber, Chronergy escapement, Glucydur balance wheel with Microstella regulating nuts; power reserve 70 hours; certified chronometer (COSC)
Functions: Hours, minutes, center seconds
Case: Stainless steel; 41 mm diameter, 11.6 mm thickness; sapphire crystal; screw-down crown; waterproof to 10 bar
Bracelet: Oyster stainless steel, folding clasp
Price: $5,975

Oyster Perpetual Datejust 41

Reference: 126300
Movement: Self-winding, Rolex caliber 3235; 29.1 mm diameter; 31 jewels; 28,800 vph; Parachrom hairspring, Paraflex shock absorber, Chronergy escapement, Glucydur balance wheel with Microstella regulating nuts; power reserve 70 hours; certified chronometer (COSC)
Functions: Hours, minutes, center seconds; date
Case: Stainless steel; 41 mm diameter, 11.6 mm thickness; sapphire crystal; screw-down crown; waterproof to 10 bar
Bracelet: Jubilee in stainless steel, folding clasp, with extension link
Price: $8,041

Oyster Perpetual Datejust 41

Reference: 126334
Movement: Self-winding, Rolex caliber 3235; 29.1 mm diameter; 31 jewels; 28,800 vph; Parachrom hairspring, Paraflex shock absorber, Chronergy escapement, Glucydur balance wheel with Microstella regulating nuts; power reserve 70 hours; certified chronometer (COSC)
Functions: Hours, minutes, center seconds; date
Case: Stainless steel; 41 mm diameter, 11.6 mm thickness; bezel in white gold; sapphire crystal; screw-down crown; waterproof to 10 bar
Bracelet: Oyster stainless steel, folding clasp, with extension link
Price: $9,828

Oyster Perpetual Datejust 41

Reference: 126334
Movement: Self-winding, Rolex caliber 3235; 29.1 mm diameter; 31 jewels; 28,800 vph; Parachrom hairspring, Paraflex shock absorber, Chronergy escapement, Glucydur balance wheel with Microstella regulating nuts; power reserve 70 hours; certified chronometer (COSC)
Functions: Hours, minutes, center seconds; date
Case: Stainless steel; 41 mm diameter, 11.6 mm thickness; bezel in white gold; sapphire crystal; screw-down crown; waterproof to 10 bar
Bracelet: Jubilee in stainless steel, folding clasp, with extension link
Price: $10,051

Oyster Perpetual Datejust 36

Reference: 126233
Movement: Self-winding, Rolex caliber 3235; 29.1 mm diameter; 31 jewels; 28,800 vph; Parachrom hairspring, Paraflex shock absorber, Chronergy escapement, Glucydur balance wheel with Microstella regulating nuts; power reserve 70 hours; certified chronometer (COSC)
Functions: Hours, minutes, center seconds; date
Case: Stainless steel; 36 mm diameter, 11.3 mm thickness; yellow gold bezel; sapphire crystal; screw-down crown, in yellow gold; waterproof to 10 bar
Bracelet: Oyster stainless steel with yellow gold elements, folding clasp, with extension link
Price: $11,224

Oyster Perpetual Datejust 31

Reference: 278384RBR
Movement: Self-winding, Rolex caliber 2236; 20 mm diameter, 5.95 mm height; 31 jewels; 28,800 vph; Syloxi hairspring, Glucydur balance wheel with Microstella regulating nuts; power reserve 55 hours; certified chronometer (COSC)
Functions: Hours, minutes, center seconds; date
Case: Stainless steel; 31 mm diameter, 11 mm thickness; white gold bezel set with 46 diamonds; sapphire crystal; screw-down crown; waterproof to 10 bar
Bracelet: Oyster stainless steel, hidden folding clasp
Remarks: Dial with diamond "VI" numeral.
Price: $16,361

Cellini Moonphase

Reference: 50535
Movement: Self-winding, Rolex caliber 3195; 28.5 mm diameter; 31 jewels; 28,800 vph; Parachrom hairspring, Paraflex shock absorber; power reserve 48 hours; certified chronometer (COSC)
Functions: Hours, minutes, center seconds; date, moon phase
Case: Rose gold; 39 mm diameter; sapphire crystal; screw-down crown; waterproof to 5 bar
Strap: Reptile leather, folding clasp
Remarks: Blue-enameled sky disc with moon made of rhodium-plated meteorite
Price: $27,250

Cellini Time

Reference: 50509
Movement: Self-winding, Rolex caliber 3132 (basic caliber 3130); 28.5 mm diameter; 31 jewels; 28,800 vph; Parachrom Breguet hairspring; power reserve 48 hours; certified chronometer (COSC)
Functions: Hours, minutes, center seconds
Case: White gold; 39 mm diameter; sapphire crystal; screw-down crown; waterproof to 5 bar
Strap: Reptile leather, pin buckle
Price: $15,468
Other versions: In rose gold ($15,468)

Cellini Time

Reference: 50505
Movement: Self-winding, Rolex caliber 3132 (basic caliber 3130); 28.5 mm diameter; 31 jewels; 28,800 vph; Parachrom Breguet hairspring; power reserve 48 hours; certified chronometer (COSC)
Functions: Hours, minutes, center seconds
Case: Rose gold; 39 mm diameter; bezel set with diamonds; sapphire crystal; screw-down crown; waterproof to 5 bar
Strap: Reptile leather, pin buckle
Remarks: Dial with diamond indices
Price: On request

Cellini Date

Reference: 50519
Movement: Self-winding, Rolex caliber 3165 (Rolex basic caliber 3187); 28.5 mm diameter; 31 jewels; 28,800 vph; Parachrom-Breguet hairspring; power reserve 48 hours; certified chronometer (COSC)
Functions: Hours, minutes, center seconds; date
Case: White gold; 39 mm diameter; sapphire crystal; screw-down crown; waterproof to 5 bar
Strap: Reptile leather, pin buckle
Price: $18,148
Other versions: A range of dial designs; in rose gold

Cellini Dual Time

Reference: 50529
Movement: Self-winding, Rolex caliber 3180 (Rolex basic caliber 3187 with module); 28.5 mm diameter; 31 jewels; 28,800 vph; Parachrom hairspring; power reserve 48 hours; certified chronometer (COSC)
Functions: Hours, minutes, center seconds; additional 12-hour display (second time zone), day/night indicators
Case: White gold; 39 mm diameter; sapphire crystal; screw-down crown; waterproof to 5 bar
Strap: Reptile leather, pin buckle
Price: $19,712
Other versions: A range of dial designs

Cellini Dual Time

Reference: 50525
Movement: Self-winding, Rolex caliber 3180 (Rolex basic caliber 3187 with module); 28.5 mm diameter; 31 jewels; 28,800 vph; Parachrom hairspring; power reserve 48 hours; certified chronometer (COSC)
Functions: Hours, minutes, center seconds; additional 12-hour display (second time zone), day/night indicators
Case: Rose gold; 39 mm diameter; sapphire crystal; screw-down crown; waterproof to 5 bar
Strap: Reptile leather, pin buckle
Price: $19,712
Other versions: A range of dial designs

Other Schiffer Books on Related Subjects:

The Rolex Story
Franz-Christoph Heel, ed.
ISBN 978-0-7643-4597-5

Vintage Rolex Sports Models, 4th Edition: A Complete Visual Reference & Unauthorized History
Martin Skeet & Nick Urul
ISBN 978-0-7643-5844-9

Rolex Highlights
Herbert James
ISBN 978-0-7643-4684-2

Originally published as *Armband Uhren Special: Rolex*, ©2020: HEEL Verlag GmbH, Königswinter
Translated from the German by Simulingua, Inc.

Library of Congress Control Number: 2022932046

Editor: Franz-Christoph Heel
Publishing manager for magazines: Sabine Blüm

Responsible for the content:
Peter Braun (editor in chief, Armbanduhren [Wristwatches] magazine).
Editorial address: Friedrichsplatz 12, 68165 Mannheim, Germany.
Telephone: 0621 712202; p.braun@heel-verlag.de.

Authors: Peter Braun, Martin Häussermann, Dr. Harry Niemann, Iris Wimmer-Olbort

Photographers: Rolex, Dr. Crott Auctioneers, Jörg Hajt, Martin Häussermann

Editing: Melanie Jaschob-Ahaus

Typesetting and layout: Axel Mertens, HEEL Verlag GmbH

Cover design: Christopher Bower

Advertising manager: Sabine Blüm

Type set in Bodoni URW/Tisa Pro

ISBN: 978-0-7643-6453-2
Printed in India

Published by Schiffer Publishing, Ltd.
4880 Lower Valley Road
Atglen, PA 19310
Phone: (610) 593-1777; Fax: (610) 593-2002
Email: Info@schifferbooks.com
Web: www.schifferbooks.com

SUPERLATIVE CHRONOMETER
OFFICIALLY CERTIFIED